Dr Richard H.

School of Pharmacy

University of California

San Francisco

Statistics:
A Conceptual Approach

Statistics: A Conceptual Approach

Sidney J. Armore

Charles E. Merrill Publishing Company
A Bell & Howell Company
Columbus, Ohio

This book was set in 10/12 Times Roman
The production editor was Joanne Morelli Harris
The cover was designed by Will Chenoweth

Published by
Charles E. Merrill Publishing Company
A Bell & Howell Company
Columbus, Ohio 43216

International Standard Book Number: 0–675–08730–9

Library of Congress Catalog Card Number: 74–22892

1 2 3 4 5 6 7 8 9 10 —

Portions of this book adapted from Sidney J. Armore, *Introduction to Statistical Analysis and Inference for Psychology and Education,* © 1966 by John Wiley & Sons, Inc. by permission of John Wiley & Sons, Inc.

Printed in the United States of America

To my wife, ETHEL, and to the memory of my father, BENJAMIN ARMORE, who instilled in his children a desire to pursue education and learning and whose energy, drive, and perseverance inspired them to reach out toward worthwhile goals.

Preface

This book was written for nonstatistics majors and other nonstatisticians who use statistical data and encounter statistical concepts in their work and reading.

The organization and structure of this text represents an innovation, as explained in the note to instructors, which follows the preface. The book is not mathematical in any way, except for the technical notes which appear at the end of the chapters. However, some very simple arithmetic is presented in a few places (involving only addition, subtraction, division, and multiplication) to enrich the readers' understanding of the concepts. This book presents statistical methods, terminology, and concepts and the basis for statistical thinking in an easy-to-understand, completely nonmathematical style. The problems discussed are taken from real-life examples.

Technical aspects of the methods presented are explained in a simplified yet adequate manner in the technical notes. Statistical methods for small samples and nonparametric methods are included in Appendix A. The technical notes and appendices are entirely supplementary and are optional.

Special emphasis is given to the statistical thinking which underlies the methods used by statisticians. The text helps nonstatisticians (1) to understand, evaluate, and use the reports of professional statisticians as he encounters these in his work; (2) to better communicate with the statistician when discussing statistical projects to be carried out for his use or the findings of a statistical analysis; (3) to gain an appreciation of what statistics can and cannot do; (4) to obtain a general understanding of the scientific method of inductive inference as applied to statistical problems.

I am indebted to the faculty collegues and other associates who have contributed in various ways to my thinking on how an introductory statistics book should be constructed. I am thankful for the contribution of the students in my statistics classes and for the understanding they gave me of the problems they encounter in learning statistics. I am especially grateful to Professor Wolfe of the Ohio State University and Professor Charles Locke of the University of South Carolina for their helpful comments.

In a special way, I am indebted to my wife, Ethel Armore, for her understanding and assistance during the many long hours I spent writing this book. She contributed valuable ideas relating to the production aspects of textbook writing, and also typed the manuscript and proofread. My son, Jerome L. Armore, and my daughter, Susan V. Armore, contributed in the various phases of problem construction, computation, and proofreading.

Note to Instructors

Teaching methods and textbooks for introductory statistics for nonstatistics majors are undergoing intensive review and re-appraisal, as evidenced by the variety of introductory statistics courses offered in colleges. Clearly, the use of statistical methods in business, industry, government, and the professions is widespread and workers in nearly every field are faced with the need to understand at least how the statistician thinks, what his products are, what they mean, and how they are to be used. A major concern, particularly of teachers of statistics, is the failure of many introductory statistics textbooks to reach those students whose primary interest is in another area of study and who take only one statistic course. Sometimes, such students complete an introductory statistics course with little real understanding of the basic methods of statistics and the statistical thinking which underlies these methods. This book has been prepared in the hope that it will provide a text that nonstatistics majors will find interesting—one they can relate to and can learn from effectively.

The plan and organization of this text represents an innovation which, it is hoped, will succeed in reaching the nonstatistics major. The main text, which does not include the technical notes appearing at the end of a chapter, has been presented with nearly no mathematics and no mathematical symbols. This is not a skimming-of-the-surface-type of presentation. The student is provided with a firm, meaningful, substantive, and enriched understanding of statistics. The main text omits much of "hands on" statistics, since it says little about *how* to conduct statistical analysis, but concentrates on *what is done* and *what it means.*

The expectation is that after studying a chapter in the main text, the student will feel comfortable in his understanding of the topic discussed and will lose much or all of his fear of studying statistics. He will then more easily enter into the study of the technical notes at the end of the chapter. The text was constructed to be suitable for a variety of statistics courses, as follows:

1. *The standard type of introductory statistics course.* The structure of this text is particularly suited for such courses because it permits the student to grasp the meaning of a topic without the distracting and inhibiting effects of mathematical symbols and discussions. It should then be easier for the student to learn the more technical material in the technical notes. Such statistics courses may be structured to include chapters 1 through 16, chapters 19 through 22, plus a selection of topics from chapters 17, 18, and Appendix A. Technical notes related to the topics covered should be included.

2. *Very elementary statistics courses.* Such a course may include all of the main text (chapters 1 through 22) plus a selection of the material in the technical notes and Appendix A. For example, technical notes 5A, 9A, 13A, 13B, 13C, 14A, 14B, 14C, 16A, 16B, 20A, and 21A could be included, plus some of the material in other technical notes and Appendix A.
3. *Courses on understanding statistics, statistical concepts and terminology, and statistics appreciation.* These are courses that emphasize statistical thinking and the interpretation of the statistician's products. Such a course would include the main text (chapters 1 through 22), excluding the technical notes and Appendix A. An instructor may wish to supplement such a course by including some or all of the following: technical notes 5A, 9A, 13A, 20A, 21A and sections A.4 and 4.7 in Appendix A.

A generous number of problems are provided for instructor and student use. These are in two parts, as follows:

1. The problems at the end of a chapter and Appendix A relate to the discussions in the main text. These include comment and brief discussion-type problems that are meant to encourage the student to probe the text and, perhaps, study more carefully. Answers to even-numbered problems are in Appendix C.
2. The instructor's manual contains additional problems relating to the main text (many of which enrich and extend the discussions in the text) and a full set of problems relating to the technical notes. Instructors will find it convenient to use these problems for quizzes and examinations. No doubt, some of the problems relating to the technical notes will be assigned as homework exercises to provide the student with study material relating to these notes. Instructors who wish to make these problems available to the students may arrange with the publisher to make the instructor's manual available for student use. This manual also contains answers to the odd-numbered problems in the text and answers to the problems in the instructor's manual.

Contents

Part D Sample Data in Population Estimates

Part E Association and Prediction

Appendix A. Additional Methods

Part A

STATISTICS TODAY: USEFULNESS AND GENERAL CONCEPTS

1

The Nature of Statistics

1.1 The Meaning of Statistics

What is *statistics*? The popular meaning refers to a set of numerical data or files of numerical records. For example, "crime statistics" refers to the numbers of arrests, felonies, etc., reported by the police department. "Vital statistics" refers to the numbers of births, deaths, marriages, etc., collected and reported by municipalities. The production records of automobile manufacturers, daily quotations of wholesale beef prices, and the list of IQ scores for freshmen entering a university are all examples of statistics.

Statistical methods are a more technical aspect of statistics. These methods include a wide variety of techniques for collecting, organizing, tabulating, and analyzing numerical data as well as for interpreting, describing, and presenting the data. For example, statistical methods are used when data on the numbers of employed women are collected from manufacturing firms, when collections of data on the weekly production of trucks are summarized into useful tabulations, when school children are given reading-readiness tests to evaluate their reading ability, and when voter preferences are determined and used to predict the outcome of an election.

A single item of data is called a *statistic*, for example, the number of divorces during a given month in a city, the price of a particular stock on a certain day, the batting average for a particular player, or the number of new one-family homes sold during a week. "Statistic" has an important technical meaning as well. It is a computed measure, such as an average or a percentage, based on a sample.

1.2 Usefulness of Statistics

There is no restriction on the kind and quantity of numerical data to which statistical methods can be applied. Suppose the thicknesses of several plastic sheets produced by a machine are to be recorded and analyzed in order to evaluate the precision of the machine. Statistical methods are available for this purpose. Suppose monthly sales and prices are to be tabulated and analyzed by a manufacturer to determine the effect of price changes on sales, or it is desired to evaluate the attitude of consumers toward a product by collecting information from a sample of consumers. Statistical methods are available for these purposes.

When a problem needs to be solved, a decision needs to be made, or a question needs to be answered, it is not always necessary to employ statistical methods. However, when the need to use statistical method arises, it is always because there is a problem to be solved, a decision to be made, or a question to be answered. For example, a company collects data on employee ages, by occupational categories, to determine the advisability of establishing a training program. Data are collected and analyzed relating to consumer acceptability of a new product to aid the producer in making final plans for mass production. A school psychologist administers psychological tests to a sample of pupils to determine the effects of poverty on school performance.

Generally, a most important part of the application of statistical methods is to carefully and precisely define the problem to be solved, the decision to be made, or the question to be answered.

2

Statistics Is for Nearly Everyone

2.1 Statistical Methods Are Versatile

Statistical methods have very wide applicability. They are used in every phase of business activity, including production, management, marketing, investment, advertising, accounting, and economic and financial analysis. Statistical methods are used extensively in public administration, education, psychology, sociology, medicine, agriculture, engineering, etc. The same basic methods are used in all these areas.

A knowledge of statistical concepts and terminology is vitally important to the progressive man and woman in business and the professions, and it will become even more important as technology advances and our economy grows in size and complexity.

Historically, statistics used in business involved, largely, data collection and data description. For example, a firm would compile data on daily, weekly, or monthly production, perhaps by type of product. Based on these records, production trend charts would be plotted to show production over time.

In another example, a distributor would keep close records of sales, by commodity, geographical region, and, perhaps, other details. Various tabulations and graphs of the data would be made to show sales growth by commodity and region. Regional or, better yet, city sales volume, would be indicated by sticking pins with different colored heads into a wall map (each color indicating a different sales volume) to display the market distribution of company sales.

The historical use of statistics, as just illustrated, has been to describe events of past and current periods and to provide a basis for thinking and planning ahead, usually on an intuitive and judgmental basis. However, in recent times there has been considerable development of statistical theory and methods, shifting the major emphasis and usefulness of statistical methods into the areas of sample survey design and estimation, quality control, forecasting, and hypothesis testing.

2.2 Illustrative Applications of Statistics

Statistical methods provide an objective, carefully structured approach to the ever more complex and demanding task of decision making. Let us consider several situations which illustrate the need for statistical decision making.

Example 1 A food manufacturer is planning the production and marketing of a new type of breakfast food. Answers must be found to various questions before final plans can be made. For example, it is thought that teenagers may constitute a major market for this product. It is important to make a reliable determination of this possibility because the product could be manufactured in a variety of flavors and certain flavors may be more popular in the teenage market than others. Furthermore, the packaging form and design will need to take into consideration the major consumers of the product. Other considerations, too, will need to be taken into account. Should the product be marketed in tablet, powder, or liquid form; in large, small, or medium size units; with sweetening already added or to be added by the consumer? Some of the answers may be provided by company experts or outside consultants. However, the answer to the basic question of whether teenagers may be expected to make up a major part of the market for the new type of breakfast food must be determined by a market survey carefully designed and evaluated by appropriate statistical methods. Some of the other answers required may likewise come from the same or additional surveys.

Example 2 An East Coast department store chain wishes to expand by establishing an outlet in the Midwest. Several possible locations are under consideration. Each location must be studied to provide a basis for selecting the one with the best profit potential. It is necessary to collect and study in each location data on population and income distributions and trends, tastes and preferences, rates of family formation, migration patterns and rates, and housing characteristics (proportion of residents who are homeowners, rent levels, residential real estate prices, etc.). These and other economic factors must be projected into the future, since a department store is a long-term operation. These studies will be used to determine the most desirable location, the initial size of the store to be constructed, the type and size of merchandise inventory, and the potential for future growth. Statistical methods are available to plan and conduct such studies and to evaluate the findings.

Example 3 A metal processing firm purchases a number of special parts from several producers. It is necessary that certain of the parts meet rigid specifications determined by company engineers. If specifications are not met within stated tolerances, production delays will occur and costs will rise. This is considered a serious problem since the firm works on a close profit margin, due to the intense competition it faces from other companies. In order to protect itself, the firm must insure that the large volume of parts purchased meets the specifications. Clearly, it would be unreasonable and costly to have company employees inspect every part before it is used in production. The usual procedure is for the firm to use a *sample inspection plan* to provide the needed protection. This may be accomplished by the application of appropriate statistical methods of *quality control.*

Other examples A sociologist is appointed to head a committee to study the pattern and incidence of crimes in a city to determine the major factors related to an increase in the crime rate. A team of experts analyzes and evaluates the characteristics of the poor and the incidence of poverty in order to develop a program to eliminate poverty. A team of medical experts analyzes the costs of health care and the distribution and effectiveness of medical services in order to recommend a national program to improve health care. Statistical methods are available which are appropriate for such projects.

Review

Concepts and Terminology

Estimation
Forecasting
Hypothesis testing
Sample survey design
Statistical decision making
Quality control

Problems

1. "Nonstatisticians have no need to know about statistical concepts and methods and need to know very little about statistical terminology. We have statisticians to do this work for us." Comment briefly.
2. "To me, statistics means a collection of figures. For example, vital statistics is a collection of figures showing the number of births, deaths, marriages, and divorces each month." Comment briefly.
3. The police commissioner commented before a group of concerned citizens, "There are certain things each of you could do to keep from becoming a crime statistic." What did he mean by "becoming a crime statistic"?
4. Define and illustrate: statistics (popular meaning), statistics (technical meaning), statistic (two meanings).
5. "We all know that statistical methods can be applied only when you have a large quantity of data." Comment briefly.
6. Statistical methods are not always useful when a problem needs to be solved, a question needs to be answered, or a decision needs to be made. Discuss briefly.

7. Suggest a purpose or purposes for each of the following compilations of data:
   (a) A shoe-store owner records the number of pairs of ladies shoes sold by color and style for a period of weeks.
   (b) A retail firm records, for each sales clerk, the percentage of weekly items sold which are returned by the purchasers because of poor fit.
   (c) The plant purchasing manager requests his staff to determine the weight per fitting for a sample of fittings purchased from a new supplier.
   (d) A medical research team determines the weights for a sample of men who have a certain disease, both before and after a new type of treatment.
   (e) A soil scientist plants 10 seeds of a given plant variety in each of 20 plots, where the plots contain different types of soil. After a period of growth, he determines the average plant height per plot and prepares a tabulation of these data.

3

Populations and Samples

The concepts of a *population* and a *sample* are fundamental in statistics. Generally, statistics deals with two types of data, population data and sample data.

The complete set of data relating to an area of interest is called the *population.* A job analyst planning to make a study of the weekly salaries paid to the 10,500 computer programmers in a city is interested in the population of 10,500 weekly salaries of computer programmers. As a matter of convenience, this may be loosely referred to as a population of computer programmers, even though the interest is the weekly salaries paid to these programmers. A population is also called a *universe.*

A residential builder may wish to study the family-income distribution in a city. If there are 50,000 families in the city, he is interested in the population of 50,000 family incomes (or 50,000 families). Suppose this population includes 5,000 high-income families, 35,000 middle-income families, and 10,000 low-income families, and the builder is interested in these three family-income groups separately, not combined. From his point of interest, there are three family-income populations: a high-income, a medium-income, and a low-income population.

It is more likely, though, that the builder would be interested in all family incomes as a group and also in each of the three categories separately. In that event, the builder would be interested in a single population (the population of all family incomes in the city) and in three *subpopulations* (each of the three family-income categories).

A set of data represents a *sample* if it contains less than the total amount of information in an area of interest. For example, a sociologist wishes to make a study of the number of serious crimes committed last year in cities with a population of 50,000 or more, but collects data from only 100 out of the large number of such cities in the country. His data on the counts (numbers) of serious crimes constitute a sample of 100 counts selected from the large population of counts which could have been collected.

In another example, a security analyst plans to study daily price relationships among a group of 20 stocks for a five-year period, but obtains price data only for Tuesdays and Fridays. The Tuesday and Friday prices constitute a sample of the total population of daily prices which could be collected for the five-year period for the 20 stocks. Clearly, a sample represents a part of a population.

3.1 Statistical Methods for Population Data

Statistical methods appropriate for use with population data are called descriptive statistics, since the function of such methods is to describe the population. These methods are also useful in describing sample data. More specifically, descriptive statistics comprise methods for organization, summarization, and analysis of data.

Suppose a school system has collected detailed information on each of its 800 administrative and clerical employees, showing length of employment in the school system, age, salary, date of last change in salary, etc. It is impossible to learn much from such a large volume of data without some kind of organization and summarization. The data may be organized into tabulations showing the numbers of employees in various job classifications, age groups, and salary categories. It may be useful to compute certain descriptive measures, such as average age, average salary, and proportion of employees who had a change in salary during the past year. Descriptive statistics provide methods for organization and evaluation of such data.

3.2 Statistical Methods for Sample Data

Usually, however, statistics deals with sample data, not population data, because collection and analysis of data for an entire population is usually expensive and time-consuming, and sometimes impossible. Required information can be obtained much more rapidly (and more accurately) from a sample than from the entire population. Application of statistical methods to sample data can provide useful and adequate estimates of population characteristics. Hence, it is usually unnecessary to collect information for an entire population.

Suppose it is desired to determine the average weight of the yolks of eggs produced by a breed of chicken. The population of egg yolks is inaccessible since it is not possible to collect all the eggs laid by the given breed. Moreover, even if all the eggs could be collected, the yolks could be obtained for weighing only by breaking open the eggs, thus leaving none for sale. The only practical alternative for such populations is to estimate the average based on sample data.

Sample data are collected in order to estimate population characteristics. Estimates based on sample data are *inductive inferences*, since reasoning from the particular (the sample) to the general (the population) is involved. Statisticians speak of this as *generalizing from the sample to the population.*

Suppose a sample of brass fittings selected from a large supply shows that 4% are damaged. This finding is generalized to the entire population by inferring (estimating) that 4% of the entire supply of fittings are damaged. Statistical methods applied to sample data to make inferences about the population are called *inductive statistics.*

In statistics, inductive inference is called *statistical inference.* Risk is always involved when an inference is made. For example, a new employee may make many mistakes in his work and his supervisor is likely to make the inference that this employee is inefficient. However, the employee may be so anxious to do good work and hold his new job that his usually efficient work habits are impeded by excessive nervousness.

Similarly, risk is always present when a statistical inference is made. Suppose a sample of sales slips in a large department store shows that 20% are for cash sales. The inference is then made that 20% of all the sales slips are for cash sales. How good is this estimate (inference)? It is hardly to be expected that this estimate is exactly right. However, is 20% just about right or is it way off?

How do statisticians determine the risks involved in making a statistical inference? Inductive statistics, based on the mathematics of *probability*, provide methods for making statistical inferences about a population based on sample data and for evaluating the risks involved. More will be said about this later.

The reader should note that all statistical methods are directed toward description of a population. Descriptive statistics are used to describe a population based on population data. Inductive statistics are used to describe a population based on sample data.

Review

Concepts and Terminology

Descriptive statistics
Generalizing from the sample to the population
Inductive inference
Inductive statistics
Population
Probability
Risk
Sample
Statistical inference
Subpopulation
Universe

Problems

1. Define and illustrate: population, sample, subpopulation, universe.
2. In each case identify the population, subpopulation, and/or sample as appropriate:
   (a) An accountant wishes to determine how many of a firm's delinquent accounts have a balance due exceeding $1,000.

(b) A traffic engineer is asked to determine which of the major intersections in a city should have a policeman directing traffic in addition to traffic lights.
(c) A high school teacher selects the IQ scores for 25 students from the senior students' IQ scores to obtain an understanding of how these scores for seniors are distributed.
(d) A drug manufacturer plans to study the reaction time of 30 men and 30 women to a new drug to find out what the reaction time is for adults in the 40- to 50-year age group and whether the reaction time differs between men and women in this age bracket.

3. A business analyst selects a sample of 50 small firms to determine the unit cost of production for smal! firms in a given industry. Would you say that the population of interest here is "the population of small firms in the industry?" Why?

4. Use an illustration to explain the meaning of: descriptive statistics, inductive statistics.

5. Each household in a city was asked how much was spent for food during the previous week. Which statistical methods are appropriate to summarize and analyze these data?

6. A sample of married men was selected from the population in a region and each man in the sample was asked how much life insurance he carried. If these data are used to estimate the proportion of the married men in the population who have more than $10,000 life insurance, which statistical methods are being used?

7. A research biologist is interested in the effectiveness of a new chemical in preventing apple scab. He selects a sample of apples collected from an orchard of 200 trees to estimate the percentage of apples in the orchard affected by apple scab. The apple trees in the orchard have been treated with the new chemical during the growing season.
(a) Identify the population involved in this problem.
(b) Which area of statistical methods is involved?
(c) What is the risk involved in this problem?

8. What does the statistician mean when he states that he is generalizing from a sample to the population?

4

Constants, Variables, and Scales of Measurement

It is frequently the practice of manufacturers of consumer products to maintain a fixed retail price for a commodity as long as it is feasible to do so. Such a price, during the period it remains unchanged, is a constant price. If a machine is set to produce at a fixed rate of 100 fittings per hour, this machine has a constant rate of production. A *constant* is a quantity with a fixed value.

4.1 Variables

A *variable*, on the other hand, is a quantity which can take on different values and is a basic concept in statistics. The daily wholesale price of eggs in a city is a variable, since it may be expected to vary from day to day. Common stock prices which appear in the financial section of a newspaper represent a variable, since price varies from stock to stock and, for each stock, from day to day.

Consider such variables as the number of accidents per week in an industry, the number of children per family, and the number of rainy days per month. These are *discrete variables* because they can take on only specified values. For example, the number of children per family can be only a whole number such as 0, 1, 2, etc. Variables which result from counting are always discrete. Generally, variables restricted to certain values are discrete, such as the variable employee salary paid last week by a firm which paid only the three amounts of \$100, \$125, and \$150.

Variables which result from measuring (other than counting) are conceptually *continuous variables*. Some variables of this type are length, time, and weight. Such variables can take on *any* value. For instance, height can equal 12, 13, or 14 inches or any in-between value such as 13.5, 13.81, or 13.372 inches.

4.2 Scales of Measurement

The values which a variable can take on represent a *scale* which may be spoken of as a *scale of values* or a *scale of measurement*. If the variable is discrete, we have a *discrete scale*; if it is continuous, we have a *continuous scale*. The distinction between discrete and continuous scales is useful in statistics.

There is yet another classification of scales which should be noted. This is the classification according to *level of measurement*. Generally, statisticians recognize four scales (or levels of measurement): *nominal scale*, *ordinal scale*, *interval scale*, and *ratio scale*. These scales differ with respect to the amount of measurement information they provide.

The *nominal scale* is a scale of classification which provides the minimum level of measurement information. Suppose employees in a firm are classified according to the nominal scale "country of birth." Employees in the same category are equivalent (equal) with respect to the characteristic measured (country of birth). A nominal scale is, then, merely a set of nonoverlapping categories. Other examples of nominal scales are sex, race, and color.

The *ordinal scale* represents the next higher level of measurement. This is a scale of rank or relative importance. For example, corporation officers may be given the titles president, vice-president, or treasurer. These titles make up an ordinal scale and contain two *degrees of measurement*—equivalence and relative importance (greater than or less than). That is, company vice-presidents are equivalent in rank and the president is higher in rank.

Another illustration of an ordinal scale is the response scale to the question: "Do you prefer soap A to soap B?" The response scale may be "greatly prefer A," "prefer A," "no preference," "prefer B," or "greatly prefer B."

If five job applicants are ranked from 1 through 5, where 1 equals the most qualified, and so on, the ranks (1, 2, 3, 4, 5) make up an ordinal scale. Such an ordinal scale is also called a *ranking scale*.

A shortcoming of ordinal scales is that they do not indicate degree of difference. Specifically, the rank-one job applicant is better qualified than the rank-two applicant, according to the foregoing ordinal scale, but we have no measure of *how much* better. The *interval scale*, the next higher level of measurement, does provide such information. An interval scale contains the three degrees of measurement—equivalence, relative importance, and a measure of how much greater or lesser in importance.

Suppose the average number of hours of overtime worked per employee is used as a measure of employment level in a new, fast-growing industry. The scale of values for this variable is an interval scale. Assume average overtime for weeks one, two, and three was 1, 4, and 10 hours. Then, it is clear that average overtime for week two was more than for week one and also that it was 3 hours more. Similarly, we know that the average for week three was 6 hours more than for week two.

The temperature scale on a Centigrade thermometer is also an interval scale (zero degrees is the freezing point of water and 100 degrees is the boiling point of water).

That is, we know that 50 degrees is more than 40 degrees and also that it is 10 degrees more than 40 degrees.

A limitation of the interval scale is the lack of an *absolute zero point*. For example, in the previous example, if "average number of hours of overtime worked per employee" is used as a measure of employment level, zero overtime is not an absolute zero point because it does not indicate zero employment, but zero *overtime* employment. On the Centigrade thermometer, zero temperature is not an absolute zero point because it does not indicate total absence of temperature. (It indicates temperature equal to water temperature at the freezing point.) In an interval scale, the location of the zero point is arbitrary and does not indicate a *true zero point*.

The highest level of measurement is contained in a *ratio scale*. This scale contains the three properties of an interval scale plus a true zero point. The scales on which we measure length, weight, and time are illustrations of ratio scales. If an object weighs zero pounds, this object has no weight. If an object is zero inches long, this object has no length. If it takes zero seconds for a reaction to occur, the reaction is instantaneous (it takes no time at all). These are illustrations of true zero points.

In some sciences special measurement problems exist. For example, a test in accounting may be scored from 0 to 100. However, this is not a ratio scale, since a score of zero does not really mean that a student does not know anything about accounting. He may know as little as one or two very simple points, which were not covered in the test, but he does know something.

Consider the three accounting test scores 70, 75, 80. The five-point difference from 70 to 75 is not likely to indicate as much additional accounting knowledge as the five point difference from 75 to 80. Since the measurements of how much more are not meaningful in such scales, they cannot be considered as true interval scales either. At best, we can consider them to be ordinal scales. Problems such as these are also to be found in psychological test scores.

Not all arithmetic (nor all statistical) operations are permissible for some scales of measurement. For example, none of the arithmetic operations are permissible for nominal scale or ordinal scale data. For example, it makes no sense to add, subtract, multiply, or divide rank data (1, 2, 3, etc.). Sometimes it is useful to assume that data represent a higher level of measurement than is really the case. For example, test scores are usually assumed to represent ratio scale data so that such measures as the average test score can be computed. As a matter of convenience, we may refer to nominal and ordinal scale data as *qualitative data* and interval and ratio scale data as *quantitative data*.

Review

Concepts and Terminology

Absolute zero point
Constant
Qualitative data
Quantitative data

Concepts and Terminology (*cont.*)

Continuous variable
Degree of measurement
Discrete variable
Equivalent
Interval scale
Level of measurement
Nominal scale
Ordinal scale
Ranking scale
Ratio scale
Relative importance
Response scale
Scale
Scale of measurement
True zero point
Variable

Problems

1. Define and illustrate: constant, variable.
2. Distinguish between and illustrate: discrete variable, continuous variable.
3. Indicate for each illustration whether a constant, discrete variable, or continuous variable is involved.
   (a) Weights of five-year-old boys
   (b) Number of errors made by typists
   (c) Average weekly price of a stock
   (d) Number of days in a week
   (e) Scores on an English composition test
   (f) Scores on a 100-item multiple choice test
   (g) Number of days in a month
   (h) Temperature at a given moment in a given place
   (i) Proportion of rainy days in a month
   (j) Diameters of the bolts produced by a certain machine
4. List the four scales of measurement and indicate the degrees of measurement each contains.
5. Indicate for each illustration whether it is true or false and briefly defend your answer.
   (a) The categories *pass* and *fail* make up a nominal scale.
   (b) The scale Brand A, Brand B, Brand C, etc., is a nominal scale.
   (c) The scale for measuring how long it takes a child to perform a specified task does not have a true zero point.
   (d) If four machines are evaluated for efficiency and A designates the most efficient, B next most efficient, etc., the scale A, B, C, D represents an interval scale.
   (e) If a scale contains a true zero point, it also has the properties of equivalence, relative importance, and degree of importance or improvement.
   (f) The ranking scale 1, 2, 3, etc. (where 1 designates the best), must be a ratio scale because we can compute the difference between ranks, and zero rank must mean no rank at all.
6. If the height of land is measured as the number of feet above sea level, what is the level of measurement?
7. It is not possible to construct a test of achievement in learning French so that the test scores represent a ratio or even an interval scale. Discuss briefly.
8. Define and illustrate: qualitative data, quantitative data.

9. When recording the weights of ten-year-old boys, it is decided to show each weight to the nearest tenth of a pound. Is the recorded variable discrete or continuous? Why?

10. One hundred rows of celery are planted in a field. An experimental chemical is sprayed over the field during the growing season. A research botanist is interested in the effectiveness of the chemical in preventing a certain viral disease from infecting the plants. He has his assistants determine the percentage of branches per plant affected by the disease (computed to one decimal place). The variable in this problem is "percentage of branches per plant affected."

(a) Is the variable continuous or discrete? Why?

(b) What is the level of measurement of the variable? Why?

Part B

STATISTICS FOR DESCRIPTION AND ANALYSIS

5

Frequency Distributions: Part I

5.1 Simple Grouping of the Data

Often we find ourselves faced with large masses of unorganized (raw) data. For example, a specialist in business trends who needs to analyze a tabulation of sales for a given month for 300 retail outlets and a social service agency director who needs to evaluate the aid given to needy families based on a tabulation of amounts paid out last month are both faced with bulks of raw data.

Statisticians have developed methods to organize, summarize, and present the important information contained in a set of raw data. Consider a specific illustration. Table 5.1 presents the total precipitation in 1970 for each of 100 cities. Even the most careful study would yield little understanding of these data. Sometimes, *ranking* the quantities (or observations)—placing them in sequence from lowest to highest—adds meaning to a set of data. However, this is useful only for very small sets. Ranking is a form of organization. What is usually required is *summarization*, as well as organization.

The *simple grouping table*, illustrated in Table 5.2, is useful as a means of data summarization. Notice that the summarization in this table permits a better "view" of the data than is possible in Table 5.1. It is evident from this table that 17 inches of total precipitation was observed for two cities and that this was the minimum precipitation. The maximum was 46 inches, observed for only one city. The largest numbers of cities (also called the largest *frequencies*) are shown between 25 and 30 inches of precipitation.

Table 5.1

Unorganized Data: Total Precipitation for 100 Cities (Inches), 1970

20	30	36	24	40
27	27	30	20	28
33	25	28	27	35
22	36	38	36	19
25	33	31	24	25
30	31	19	30	27
31	23	23	46	23
26	22	20	31	30
27	41	17	41	41
30	31	30	45	20
38	31	22	22	41
38	33	27	25	26
25	20	26	30	30
19	25	35	27	27
40	37	33	33	22
27	27	25	33	35
22	26	22	45	25
25	21	38	35	21
17	33	45	26	25
30	27	26	26	24

Table 5.2

Simple Grouping Table: Total Precipitation for 100 Cities, 1970

Precipitation (Inches)	Number of Cities	Precipitation (Inches)	Number of Cities
17	2	32	—
18	—	33	7
19	3	34	—
20	5	35	4
21	2	36	3
22	7	37	1
23	3	38	4
24	3	39	—
25	10	40	2
26	7	41	4
27	11	42	—
28	2	43	—
29	—	44	—
30	10	45	3
31	6	46	1

5.2 Frequency Distributions

A simple grouping table provides organized, detailed information and is useful for studying data in an unfamiliar area, particularly if the number of values for the variable involved is not too large. In Table 5.2, the variable "inches of precipitation" varies from 17 to 46 and has 30 distinct values. When the number of distinct values is more than about 15 or 20, summarization in a simple grouping table is usually not satisfactory for most uses. Usually, further summarization is required to assist the mind in grasping (visualizing) the distribution of the data. This is accomplished by constructing a *frequency distribution.*

Tables 5.3, 5.4, 5.5, and 5.6 present frequency distributions with varying degrees of summarization of the Table 5.1 data. In Table 5.3, the hundred observations of total precipitation are summarized into 16 *classes* or *class intervals.* Each class interval has an associated *class frequency.* The first class, 16–17 inches, has a frequency of 2, indicating that two cities fall into this class of total precipitation. Look over these four tables and observe that the data become more and more condensed (summarized) as you go from table to table. Table 5.4 has 11 classes; Table 5.5, 8 classes; and Table 5.6, 4 classes.

Table 5.3

Frequency Distribution (16 Classes): Total Precipitation for 100 Cities, 1970

Precipitation (Inches)	Number of Cities
16–17	2
18–19	3
20–21	7
22–23	10
24–25	13
26–27	18
28–29	2
30–31	16
32–33	7
34–35	4
36–37	4
38–39	4
40–41	6
42–43	—
44–45	3
46–47	1

Table 5.4

Frequency Distribution (11 Classes): Total Precipitation for 100 Cities, 1970

Precipitation (Inches)	Number of Cities
15–17	2
18–20	8
21–23	12
24–26	20
27–29	13
30–32	16
33–35	11
36–38	8
39–41	6
42–44	—
45–47	4

Notice that summarization beyond simple grouping is accompanied by some loss of information. That is, a frequency distribution indicates how many observations fit into each class interval, but provides no information as to the actual value of each observation. For example, we know from Table 5.6 that 50 cities had between 20 and

Table 5.5

Frequency Distribution (8 Classes): Total Precipitation for 100 Cities, 1970

Precipitation (Inches)	Number of Cities
16–19	5
20–23	17
24–27	31
28–31	18
32–35	11
36–39	8
40–43	6
44–47	4

Table 5.6

Frequency for Distribution for (4 Classes): Total Precipitation for 100 Cities, 1970

Precipitation (Inches)	Number of Cities
10–19	5
20–29	50
30–39	35
40–49	10

29 inches of precipitation, but there is no way of knowing from this table the actual amount of precipitation for each of these cities. Nevertheless, the gain from adequate summarization in a well-constructed frequency distribution far outweighs any loss of information. The loss is usually minor, but the gain in understanding how the data are distributed is always substantial.

A frequency distribution, with summarization beyond simple grouping, presents a compact, more easily visualized representation of the underlying distribution of the data. However, it is important to summarize just enough. If too many class intervals are used, the frequency distribution will look too "stretched out," as in Table 5.3. The human mind cannot adequately visualize a distribution if too many classes are used, and the distribution will tend to be less well defined.

On the other hand, if too few classes are used, the data are likely to be too condensed, resulting in a serious loss of information relating to the distribution of the data. Table 5.6 illustrates an overcondensed frequency distribution which is not desirable for many uses.

How much summarization of a set of data is desirable? There is no single answer to this question, since it depends entirely on why the table is constructed. If it is desired to make a detailed analysis of the data and the distribution, then the tendency would be to use many classes. Table 5.3 may be the most suitable table for certain purposes. There are times when a simple grouping table (Table 5.2) is even more suitable. The point is, how interested are you in a detailed view of the data? In some analyses, Table 5.6 (with only 4 classes) may well provide adequate information.

On the other hand, if an important use of a frequency distribution is to exhibit the underlying distribution of the data, then either extreme (too many classes or too few classes) is to be avoided. Observe Tables 5.4 and 5.5 and notice that Table 5.4 (with 11 classes) is still too long for this purpose and that Table 5.5 (with 8 classes) appears to be just about right. Notice that Table 5.5 presents a smooth, easily visualized distribution of the data.

It is necessary to have a clear understanding of the purpose to be served when a set of data is organized into a frequency distribution. Even though special requirements may need to be considered, a typically important reason for constructing a frequency distribution is to reveal the underlying distribution of the data. When constructing such a frequency distribution, the number of classes appropriate depends, to an extent, on how many observations are contained in the set of data. If the number of observations is very limited, say, about 12 or less, a simple grouping table (Table 5.2) may be adequate. Otherwise, perhaps 8–12 classes should be used. Only infrequently should as many as 15 classes be used and never more than 20 classes.

Tables 5.7 and 5.8 illustrate some pitfalls to be avoided in constructing a frequency distribution. In Table 5.7, the lower class limits (17, 21, 25, etc.) are "difficult-to-visualize" numbers. They should always be avoided. Use "easy-to-visualize" lower limits such as 3, 6, 9, . . . ; 5, 10, 15, . . . ; or 10, 20, 30, ... Such lower limits have "eye appeal" because they are easy to recognize and easy to retain. Also, multiples of such numbers may be used.

Table 5.7

Frequency Distribution with Undesirable Class Intervals

Precipitation (Inches)	Number of Cities
17–20	10
21–24	15
25–28	30
29–32	16
33–36	14
37–40	7
41–44	4
45–48	4

Table 5.8

Frequency Distribution with Overlapping Classes

Precipitation (Inches)	Number of Cities
10–20	10
20–30	60
30–40	37
40–50	10

Table 5.8 illustrates *overlapping classes*. For example, the classes "10–20" and "20–30" overlap, since they both include 20. (In such a table, the total of the class frequencies would be too large. In Table 5.8, the total comes out to 117 cities, when the data relate to only 100 cities.)

(**Note 5A,** at the end of the chapter, presents a systematic procedure for constructing a frequency distribution and for determining certain features of a frequency table.)

5.3 Tabular Analysis

Often a statistician will provide certain analytical columns of information in a frequency distribution table to give the user greater insight into the data. This is illustrated in Table 5.9 where home-to-work distances for 50 employees are presented. The *class midpoint* column is used to compensate for the fact that specific mileage information cannot be obtained from the table. For example, the table shows that home-to-work mileages for 4 employees are between 3.0 and 5.9 miles. The midpoint column indicates in a general way that "on the average" these employees travel 4.45 miles from home to work. A class midpoint represents the middle of a class interval.

The *percentage distribution* column in Table 5.9 is usually a very valuable one. For example, it may be seen at a glance that 40% of the employees travel 6.0 to 8.9 miles from home to work and 20% travel between 9.0 and 11.9 miles. Or, 40% travel, on the average, 7.45 miles (class midpoint) and 20% travel about 10.45 miles (class midpoint).

Table 5.9

Home-to-Work Distances for 50 Employees

Distance (Miles)	Number of Employees	Mid-point (Miles)	Percentage Distribution
0.0– 2.9	2	1.45	4
3.0– 5.9	4	4.45	8
6.0– 8.9	20	7.45	40
9.0–11.9	10	10.45	20
12.0–14.9	9	13.45	18
15.0–17.9	5	16.45	10
	Total 50		Total 100

Tables 5.10, 5.11, and 5.12 illustrate *cumulative frequency distributions* which are often very useful. Table 5.10 presents the "*less than*" *cumulative frequency distribution* for the home-to-work travel distances. The middle column shows the cumulative number of employees and the last column shows the cumulative percentage distribution. For example, 26 employees, or 52%, travel less than 9.0 miles from home to work and 45 employees, or 90%, travel less than 15.0 miles.

Table 5.11 presents the "*or less*" *cumulative frequency distribution*, and Table 5.12 presents the "*or more*" *cumulative frequency distribution*. The interpretation of these tables is similar to that of Table 5.10.

Table 5.10

"Less Than" Cumulative Frequency Distribution: Home-to-Work Distances for 50 Employees

Distance (Miles)	Cumulative Number of Employees	Cumulative Percentage Distribution
Less than 3.0	2	4
Less than 6.0	6	12
Less than 9.0	26	52
Less than 12.0	36	72
Less than 15.0	45	90
Less than 18.0	50	100

Table 5.11

"Or Less" Cumulative Frequency Distribution: Home-to-Work Distances for 50 Employees

Distance (Miles)	Cumulative Number of Employees	Cumulative Percentage Distribution
3.0 or less	3	6
6.0 or less	6	12
9.0 or less	28	56
12.0 or less	37	74
15.0 or less	47	94
18.0 or less	50	100

Table 5.12

"Or More" Cumulative Frequency Distribution: Home-to-Work Distances for 50 Employees

Distances (Miles)	Cumulative Number of Employees	Cumulative Percentage Distribution
0.0 or more	50	100
3.0 or more	48	96
6.0 or more	44	88
9.0 or more	24	48
12.0 or more	14	28
15.0 or more	5	10

Review

Concepts and Terminology

Classes
Class frequencies
Class interval
Class limits
Class midpoint
Cumulative frequency distribution
Frequency distribution
"Less than" cumulative frequency distribution
Observations
"Or less" cumulative frequency distribution
"Or more" cumulative frequency distribution
Overlapping classes
Percentage distribution
Ranking
Raw data
Simple grouping
Underlying distribution of a set of data
Unorganized data

Problems

1. The number of errors made by 10 coding clerks are:

Fred	47
Ed	35
Joe	27
Tim	50
Hal	32
Bob	12
Abe	23
Ike	39
Dick	16
Stan	43

Organize these data into a more meaningful tabulation (without losing any information).

2. What does the statistician mean by "raw data"?

3. An analyst has a list of 250 juvenile first-offenders listed alphabetically by surname, showing the age of each. If he is interested in determining the ages which occur most frequently in the list, would he find a relisting of the juveniles from youngest to oldest or a simple grouping table of the age data more useful? Why?

4. An accountant has an alphabetical listing of the balances outstanding for the 300 customer accounts for a firm. Would a relisting of these balances from lowest to highest be of substantial aid to the accountant in understanding how they are distributed?

5. In Problem 4, would a simple grouping table be more helpful? Why?

6. In Problem 4, what type of data organization would be most helpful to the accountant? Why?

7. Examine the simple grouping table in Table 5.2 and notice that for some precipitation amounts (such as 18 inches, 29 inches, etc.) a dash (for zero) is shown in the "Number of Cities" column. Is it useful to include such precipitation amounts in the table? Why?

8. Taking into consideration your evaluation for Problem 7, comment on the effectiveness of the following simple grouping table of the test scores for 175 students in revealing how the scores are distributed:

Scores	*Number of Students*
62	10
63	12
78	28
79	32
80	30
97	28
98	20
99	15

9. How would you define "frequency distribution"?

10. If a set of 50 measurements is organized into a frequency distribution with as many as 20 class intervals, is there loss of information? Explain.

11. Suppose you are given the data in Table 5.1 and you wish to determine how many of the 100 cities had at least 30 inches of precipitation in 1970. If this is all the information you need out of the data:

(a) Which of the four tables in the text (5.3, 5.4, 5.5, or 5.6) would be most suitable for you to construct?

(b) If you did not have to choose among the four tables noted in (a) above, indicate the kind of frequency table which would be adequate and easiest to construct.

12. A population of 214 weights (in pounds) was organized into the following frequency distribution:

Weight	*Frequency*
15–19	7
20–24	3
25–29	4
30–34	—
35–39	25
40–44	12
45–49	25
50–54	15
55–59	17
60–64	20
65–69	18
70–74	22
75–79	10
80–84	17
85–89	2
90–94	6
95–99	—
100–104	11

(a) How many class intervals are in this frequency table?

(b) How many weights are in the 45–49 pound class, 80–84 pound class, 95–99 pound class?

(c) Briefly describe the underlying distribution of the population of weights as revealed by the frequency distribution.

13. Construct a frequency distribution of nine classes for the population of weights in Problem 12 by completing the frequency column in the following table. Does this frequency distribution give a better view of the underlying distribution of the population? Why?

Weight	*Frequency*
15–24	
25–34	
35–44	
45–54	
55–64	
65–74	
75–84	
85–94	
95–104	

14. (a) Construct a frequency distribution of six classes for the population of weights in Problem 12 by completing the frequency column in the following table:

Weight	*Frequency*
15–29	
30–44	
45–59	
60–74	
75–89	
90–104	

(b) Compare the frequency distributions for the population of weights as shown in Problem 12 and as constructed in Problems 13 and 14(a). Which best exhibits the underlying distribution? Why? Describe the underlying distribution based on the one you choose.

15. Do you consider the following frequency table well constructed? Why? If you do not, explain how it can be improved.

Bolt Diameter (cm)	*Number of Bolts*
13–15	15
16–18	21
19–21	35
22–24	50
25–27	37
28–30	21

16. Do you consider the following frequency table well constructed? Why? If you do not, explain how it can be improved.

Length of Rod (Inches)	*Frequency*
40–45	5
45–50	10
50–55	20
55–60	15
60–65	8

17. The numbers of births per 1,000 population (birth rate) for a sample of cities were organized into a frequency distribution as follows:

Birth Rate	*Number of Cities*	*Class Midpoint*	*Percentage Distribution*
5.0– 9.9	2	7.95	
10.0–14.9	8	12.95	
15.0–19.9	12	17.95	
20.0–24.9	3	22.95	—
		Total	100

(a) What would you estimate as the birth rate for the 8 cities in the 10.0–14.9 class interval and also for the 12 cities in the 15.0–19.9 class interval, based on the data in this table?
(b) Complete the percentage distribution column.
(c) How many cities had birth rates of 14.9 or less?
(d) What percentage of the cities had birth rates of 15.0 or more?

18. Refer to Problem 17 and complete the following table. What percentage of the cities had birth rates of 20.0 or more?

Birth Rate	*Cumulative Number of Cities*
Less than 5.0	
Less than 10.0	
Less than 15.0	
Less than 20.0	
Less than 25.0	

19. The numbers of diseased plants per plot were organized into a simple grouping table as follows:

Number of Diseased Plants per Plot	*Frequency (Number of Plots)*
8	1
9	—
10	3
11	5
12	—
13	8
14	—
15	7
16	13
17	16
18	15
19	10
20	9
21	12
22	10
23	18
24	17
25	19

Number of Diseased Plants per Plot	*Frequency (Number of Plots)*
26	16
27	—
28	9
29	7
30	5
	Total 200

Using this information, complete the following frequency table:

Number of Diseased Plants per Plot	*Number of Plots*	*Percentage Distribution*
8–11		
12–15		
16–19		
20–23		
24–27		
28–31		

20. Using the frequency table completed in Problem 19, complete the following table:

Number of Diseased Plants per Plot	*Cumulative Number of Plots*	*Cumulative Percentage Distribution*
Less than 12		
Less than 16		
Less than 20		
Less than 24		
Less than 28		
Less than 32		

21. Using the frequency table completed in Problem 19, complete the following table (hint: Start with "28 or more" and work upwards in the table):

Number of Diseased Plants per Plot	*Cumulative Number of Plots*	*Cumulative Percentage Distribution*
8 or more		
12 or more		
16 or more		
20 or more		
24 or more		
28 or more		

Note 5A

Frequency Distribution: Construction and Analysis

Table 5A.1 presents the time required to produce one unit of production for a sample of workers. The first step in constructing a frequency distribution for a set of raw data is to determine the number of observations (data items) in the set (n), the observation with the lowest value (L), and the observation with the highest value (H). In our illustrative problem $n = 70$, $L = 12.4$ (minutes), and $H = 31.5$ (minutes). Next, compute the range (R) as $H - L$, which in our problem is $31.5 - 12.4 = 19.1$ (minutes).

Table 5A.2 may be used to determine the approximate number of class intervals to use when the purpose for constructing the frequency distribution is to exhibit the underlying distribution

Table 5A.1

Time Required to Produce One Unit for a Sample of 70 Workers (in Minutes)

13.3	19.0	24.3	18.9	12.7	22.3	21.0
24.7	25.0	15.3	25.7	16.2	26.4	24.5
17.6	28.2	16.0	22.0	26.1	22.3	24.5
19.3	22.3	22.1	17.8	13.3	20.7	28.7
19.3	25.5	12.4	18.5	15.9	25.5	28.9
28.2	23.8	16.4	27.5	21.7	21.2	14.5
14.9	25.1	25.1	20.4	15.4	28.0	30.7
24.3	28.6	28.0	25.3	18.2	32.0	31.5
30.1	30.5	31.3	30.2	28.9	31.5	32.5
27.0	27.3	23.8	31.7	26.9	28.2	18.1

Table 5A.2

Approximate Number of Classes to Use When Constructing a Frequency Distribution.

Number of Items of Data to Be Summarized	Approximate Number of Classes to Use
15– 29	5
30– 59	6
60– 99	7
100– 199	8
200– 499	9
500– 999	10
1,000– 1,999	11
2,000– 3,999	12
4,000– 7,999	13
8,000– 14,999	14
15,000– 34,999	15
35,000– 69,999	16
70,000–149,999	17
150,000–299,999	18
300,000–499,999	19
500,000 and over	20

SOURCE: Table 3.2.5 of Sidney J. Armore, *Introduction to Statistical Analysis and Inference for Psychology and Education*, © 1966, by John Wiley & Sons, Inc., New York, N.Y. Generally based on the results obtained following the procedure in H. A. Sturges, "The Choice of a Class Interval," *Journal of the American Statistical Association*, Vol. 21, 1926.

of the data. This table indicates that for $n = 70$, 7 classes should be used. Computing $19.1 \div 7 = 2.7$, we determine that if the range (19.1) is divided into 7 parts (for 7 classes), each part is of length 2.7 (minutes). Selecting a convenient (round) number close to 2.7, we choose 3.0 (minutes) as the length of class interval to use for the frequency distribution.

The next step is to determine the lower class limits. Notice that the observation with the lowest value (L) is 12.4 (minutes). Select a convenient (round) number equal to or below L as the first lower class limit for the table. Let us select 12.0 (minutes) for our problem. Then, additional lower class limits are determined by adding the length of class interval (3.0 minutes) successively, obtaining $12.0 + 3.0 = 15.0$, $15.0 + 3.0 = 18.0$, etc. Table 5A.3 presents these lower class limits in the first column. Notice that 7 lower class limits (for 7 classes) are determined.

The upper class limits are determined so as to avoid overlapping of the classes. Notice that the second lower class limit is 15.0; therefore, the upper class limit for the first class must be just below this value, 14.9. (Notice that the time measurements are in tenths of a minute.) The third lower class limit is 18.0; then, the upper class limit for the second class must be just below this value, 17.9, etc. Table 5A.3 shows the upper class limits determined in this way.

Notice that the range of values from 12.0 (lower class limit of the first class) to 32.9 (upper class limit of the seventh class) includes the full range of the seventy observations (from $L = 12.4$ to $H = 31.5$). Always check that the full range of the raw data is included before proceeding. Sometimes an additional class interval is needed to achieve this.

Table 5A.3

Time Required to Produce One Unit for a Sample of 70 Workers: Tally Table

Class Limit (Minutes)	Tally	Frequency
12.0–14.9	~~////~~ /	6
15.0–17.9	~~////~~ ///	8
18.0–20.9	~~////~~ ////	9
21.0–23.9	~~////~~ ~~////~~	10
24.0–26.9	~~////~~ ~~////~~ ~~////~~	15
27.0–29.9	~~////~~ ~~////~~ //	12
30.0–32.9	~~////~~ ~~////~~	10

The next step is to determine the number of observations which fall in each class (the class frequency). This is accomplished by tallying, as shown in the middle column of Table 5A.3. Note that the first observation in the raw data (Table 5A.1) is 13.3 and that this observation falls in the 12.0–14.9 class. Place a tally mark (a vertical stroke) in the tally column for this class. The second observation (24.7) falls in the 24.0–26.9 class. Place a tally mark in the tally column for this class, etc. The fifth tally for each class is drawn across the previous four tally marks to tie the five tally marks into a bundle. This facilitates counting the marks after all observations are tallied. Finally, count the tallies in each class and place the count in the frequency column in Table 5A.3.

Table 5A.4 shows the completed frequency distribution. Notice that *Class Limits* from Table 5A.3 are shown as "Time Required" in Table 5A.4. Similarly, instead of "Frequency," Table 5A.4 shows the column heading "Number of Workers." These are more meaningful column headings for presentation to users of the table.

Table 5A.4

Time Required to Produce One Unit for a Sample of 70 Workers

Time Required (Minutes)	Number of Workers
12.0–14.9	6
15.0–17.9	8
18.0–20.9	9
21.0–23.9	10
24.0–26.9	15
27.0–29.9	12
30.0–32.9	10

If it is desired to use a number of classes different from those suggested in Table 5A.2, divide the range of the raw data by the number of classes desired and follow the remaining steps of the foregoing procedure.

The length of class interval in the frequency table is the same for all classes, if the foregoing procedure is followed. The length of a class can easily be determined for any frequency table as the difference between successive lower class limits or as the difference between successive upper class limits. The class midpoint is computed as half the sum of the limits for a class. For example, for the first class in Table 5A.4, $12.0 + 14.9 = 26.9$ and $26.9 \div 2 = 13.45$ (midpoint).

6

Frequency Distributions: Part II

6.1 Graphic Display

Graphic display of a frequency distribution may be used to exhibit the shape of the distribution underlying the data. Three types of graphs are used for this purpose: the *frequency polygon*, the *histogram*, and the *ogive*.

Figure 6.1 presents a *frequency polygon* constructed for the home-to-work distances for 50 employees in Table 5.9 (page 26). Notice that the class midpoints are marked off on the horizontal axis and that the corresponding class frequencies are plotted as dots over the midpoints. For example, the dot above the midpoint 1.45 miles is at a height of 2 (left-scale "number of employees"), which is the corresponding class frequency. The dot above the midpoint 4.45 miles is at a height of 4, etc. The dots are connected by a line to emphasize the shape of the distribution. This graph clearly indicates that the class frequencies rise to a peak of 20 (corresponding to the 7.45-mile class midpoint) and then, generally, decline.

Sometimes it is more useful to show a *percentage scale* instead of a *frequency scale*. This is shown on the right in Figure 6.1. It is usually not necessary to show both vertical scales as in Figure 6.1, but there are occasions when this is desirable.

Figure 6.2 presents the *histogram* for the home-to-work distances. In this chart, each bar represents a class interval. The class limits which identify a class are shown beneath each bar. Both number of employees and percent are shown for each bar. However, usually only one or the other is needed. A histogram is sometimes called a *step diagram*. The area of each bar represents the class frequency, and the area of adjacent bars represents the combined frequencies of the classes. For example, the

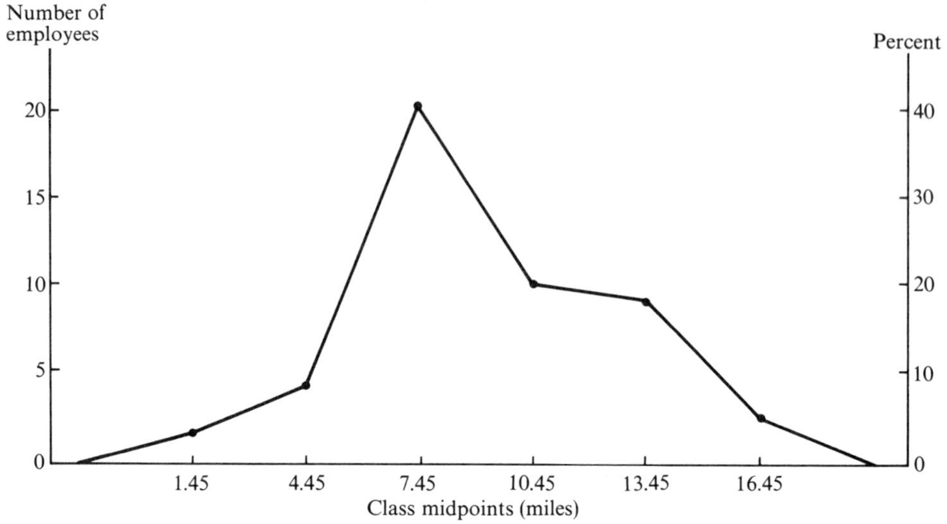

Figure 6.1

Frequency Polygon: Home-to-Work Distances for 50 Employees

area of the bar for the 0.0–2.9 mile class represents 2 employees (or 4% of the total number of employees). The combined area of the first three bars (0.0–2.9, 3.0–5.9, 6.0–8.9) represents 2 + 4 + 20 = 26 employees or 4% + 8% + 40% = 52% of the employees in the distribution. Overall, the total histogram area represents the 50 employees or 100% of the employees in the frequency distribution. The use of areas

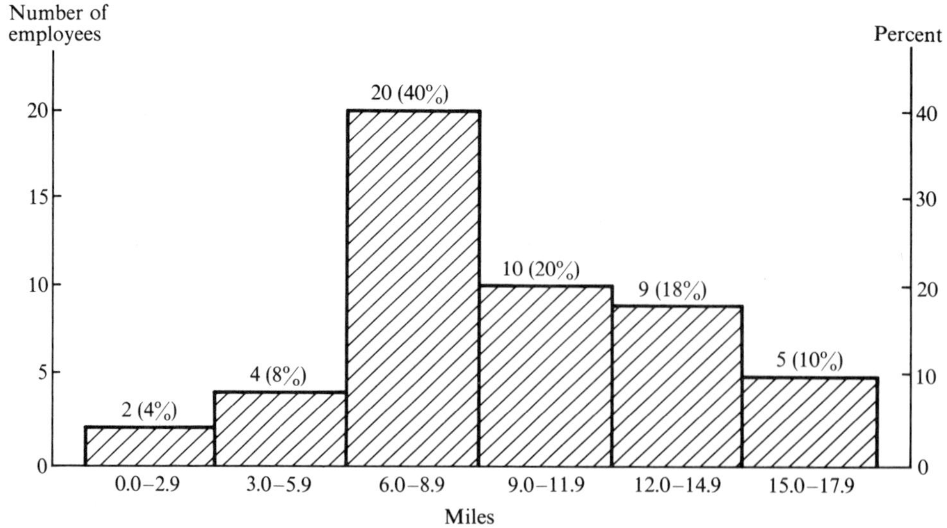

Figure 6.2

Histogram: Home-to-Work Distances for 50 Employees

under a curve, such as a histogram, to represent frequencies and especially percentages is a fundamental notion in statistics.

The *ogive* is a graph which exhibits either the cumulative frequencies or the cumulative percentage distribution. Figure 6.3 presents an ogive which displays the "less than" cumulative percentage distribution from Table 5.10 (page 27).

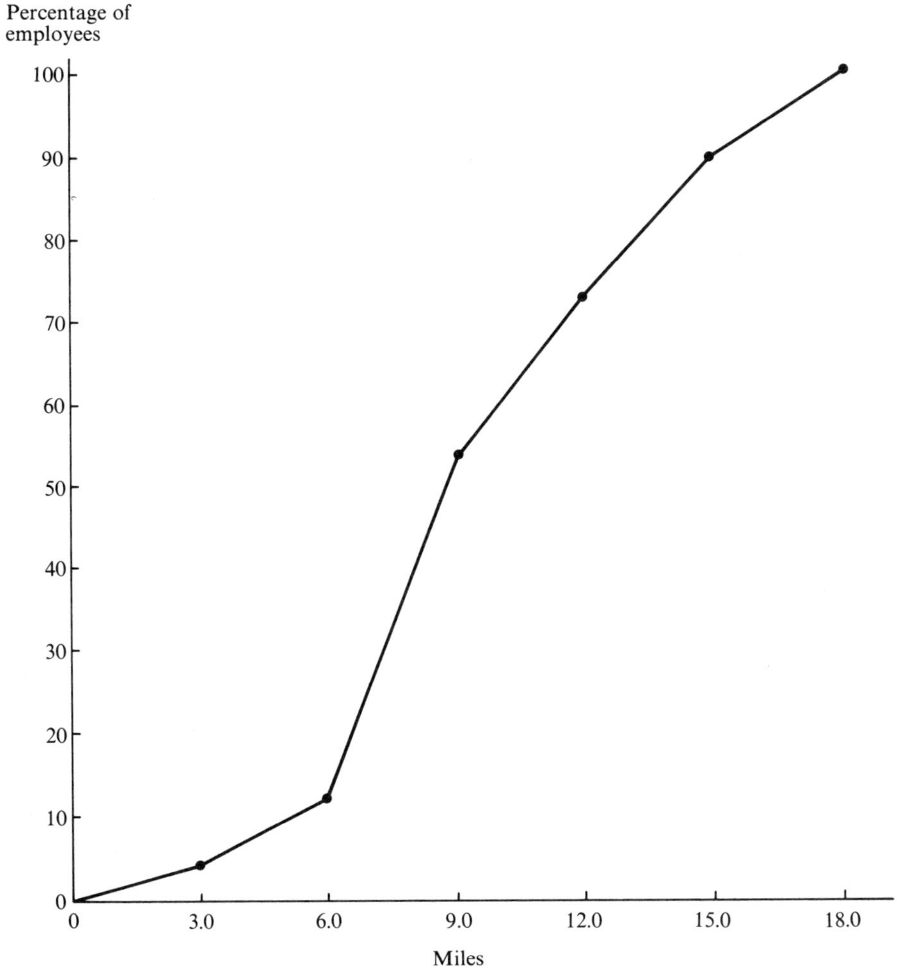

Figure 6.3

"Less Than" Cumulative Frequency Percentage Distribution: Home-to-Work Distances for 50 Employees

6.2 Class Length

The frequency distributions we have been discussing have class intervals of *equal length*. Table 6.1 presents a frequency distribution of family incomes in a city where the classes are of *unequal length*. For example, the $1,000–$1,499 income class is of length $500, since it includes a $500 range of values from the lower class limit of this class ($1,000) to the lower class limit of the next class ($1,500). The $5,000–$7,499 income class is of length $2,500, and the $10,000–$14,999 income class is of length $5,000.

Table 6.1

Family Incomes in Sundale City, 1971

Income Category	Percentage Distribution of Families
Under $1,000	3.9
$1,000–$1,499	5.2
1,500– 1,999	10.2
2,000– 2,999	14.7
3,000– 3,999	18.0
4,000– 4,999	15.4
5,000– 7,499	11.4
7,500– 9,999	9.2
10,000–14,999	8.0
15,000–24,999	2.7
25,000 and over	1.3
Total	100.0

There are situations where unequal class lengths are more convenient to use. This is often the case where a set of data includes a string of extreme values (very low or very high values). However, a frequency table with classes of unequal length may give a misleading impression of a distribution and so should be studied carefully.

For example, Table 6.1 gives the impression that larger percentages of families are concentrated in the $2,000 to $4,999 income categories than in the $1,000 to $1,999 categories. This is misleading, since the lower income categories are of length $500, whereas the higher income categories are of greater length ($1,000). Furthermore, the relatively smaller percentages of families in the highest income categories ($7,500 and up) are de-emphasized in Table 6.1 because the class lengths are so much larger. Notice that the $7,500–$9,999 class is of length $2,500; the $10,000–$14,999 class is of length $5,000, and the $15,000–$24,999 class is of length $10,000. If the family income percentages were determined for classes of smaller length, the percentages in these high income class intervals would be small indeed.

Unequal class intervals occur when *open-end classes* are used, such as "Under $1,000" or "$25,000 or over" in Table 6.1. Such classes are convenient if a few unusually high or low observations occur in a set of data. The problem with such classes is that there is no way of knowing the range of values included. For example, in the $25,000 and over class, how much over $25,000 are the incomes? It is a good idea to add a footnote to such a table to give at least the range of values included in the open-end classes. It is better if the actual values included in each open-end class are shown in the footnote.

Review

Concepts and Terminology

Area under curve
Cumulative frequency curve
Frequency polygon
Histogram
Length of class interval
Ogive
Open-end classes
Step diagram

Problems

1. In which graphic type (frequency polygon, histogram, ogive) does the area under the curve represent frequencies or percentages?
2. Why is the histogram an important type of graph from the theoretical point of view?
3. Examine Figure 6.3. Why does this curve rise as you move from left to right?
4. If the "or more" cumulative frequencies in Table 5.12 were charted in an ogive graph, would the curve increase or decrease as you move from left to right (from lower distances to higher distances)? Why?
5. Given the following frequency distribution of 175 scores on a mechanical aptitude test:

Score	*Number of Subjects*
40– 49	5
50– 59	—
60– 69	—
70– 79	2
80– 89	8
90– 99	35
100–109	50
110–119	40
120–129	15
130–139	5
140–149	9
150–159	1
160–169	—
170–179	3
180–189	—
190–199	2

(a) Reconstruct the frequency distribution by completing the following frequency table:

Score	*Number of Subjects*
40– 69	
70– 89	
90– 99	
100–109	

Score	*Number of Subjects*
110–119	
120–139	
140–159	
160–199	

(b) Does the frequency distribution just constructed have classes of equal length? What is the length of each of the 8 class intervals?

(c) Does the frequency distribution first presented have classes of equal length? What is the length of each of the 16 classes?

(d) Which of the two frequency distributions is preferable to present in a report? Why?

6. What are open-end classes? When are they useful?

7. If open-end classes are used in a frequency table, what should be done to make the table more useful to the reader?

8. A legal research assistant determined the attitude of small city legislatures concerning a bill under consideration by the Congress and constructed the following table:

City Population Size (000's)	*Number of City Legislatures in Favor of Bill*	*Class midpoint (000's)*
2.0– 3.9	5	2.95
4.0– 5.9	10	4.95
6.0– 7.9	20	6.95
8.0– 9.9	30	8.95
10.0–11.9	15	10.95
12.0–13.9	10	12.95

(a) Construct a frequency polygon.

(b) Construct a histogram. (Note: The height of a bar represents the class frequency.)

9. The Sunday supplement of a newspaper showed the following table based on a sample of 150 traffic violations. Construct an ogive.

Age of Driver	*Cumulative Number of Traffic Tickets*
16 or over	150
20 or over	120
24 or over	100
28 or over	85
32 or over	70
36 or over	55
40 or over	45
44 or over	35
48 or over	25
52 or over	20
56 or over	15
60 or over	10

7

Frequency Distributions: Part III

7.1 Distribution Shapes

Frequency distributions may be of different shapes. Figure 7.1 presents histograms for some distribution shapes which are frequently encountered. (Of course, distributions may not have exactly one of the shapes shown, but they are likely to be quite similar.) The *symmetrical distribution with central peak* is seen often. Sometimes such a distribution may be more closely concentrated (squeezed) toward the center, and the peak may be considerably higher. Or, the distribution could have a low peak and be considerably spread out at the ends.

The *skewed distribution* in Figure 7.1 has its peak off center and has one end running off into a tail. The skewed distribution illustrated has this tail on the right and is called a *positively skewed distribution.* If the tail ran off to the left, it would be called a *negatively skewed distribution.*

The *J-distribution* has its peak at one end, either the lower end (as shown in the chart) or the upper end, and is similar to a theoretical distribution called the *exponential distribution.* The *uniform distribution* has no peak and is a level distribution (all classes have the same frequency). This is also called a *rectangular distribution.*

7.2 Population and Sample Distributions

Statistical methods, whether descriptive or inductive, are directed toward description of a distribution. In the case of population data, the statistician is interested in various measures which describe the population distribution. For example, in a population

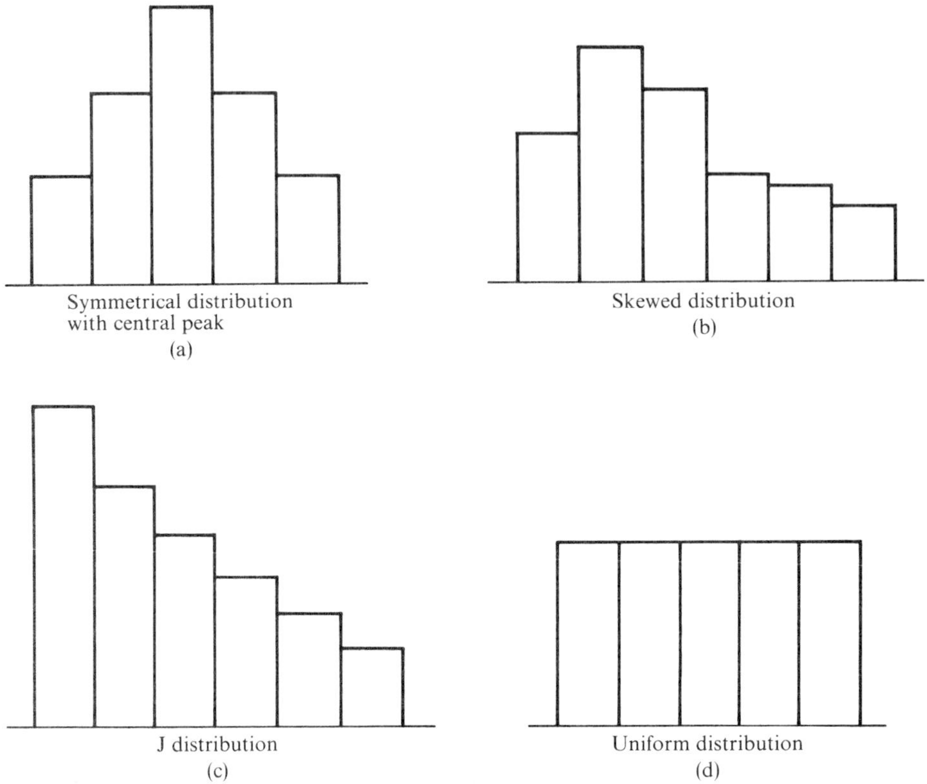

Figure 7.1

Types of Distribution

of pipes, from the statistical point of view, the interest may be in the variation in pipe diameter, the percentage of pipes defective, or the average weight per pipe. These are measures used to describe distributions.

In the case of sample data, the statistician is interested in various measures which describe the distribution of the *population from which the sample was selected.* Suppose a pollster selects a sample of voters in order to determine the proportion who favor a certain candidate. The pollster is interested in the sample information only in so far as it provides information about the entire population of voters. That is, he is interested in the proportion of the population of voters who favor the candidate.

As previously noted, different frequency distribution tables may be constructed for a set of data. However, the set of data has a specific distribution which comes through more or less sharply in a given frequency distribution table. The simple grouping table presents the true distribution for a set of population data. A bar chart is the proper form of graphic representation of a simple grouping table. Figure 7.2 illustrates such a graph for unpaid accounts in a department store, distributed according to the number of months since last payment (which was previously organized by the analyst into a simple grouping table). The bars may be of any width, as long as they are all of equal width and are placed at unit intervals. Spaces must be reserved for zero frequencies (as shown in the figure) in order not to distort the appearance of the distribution.

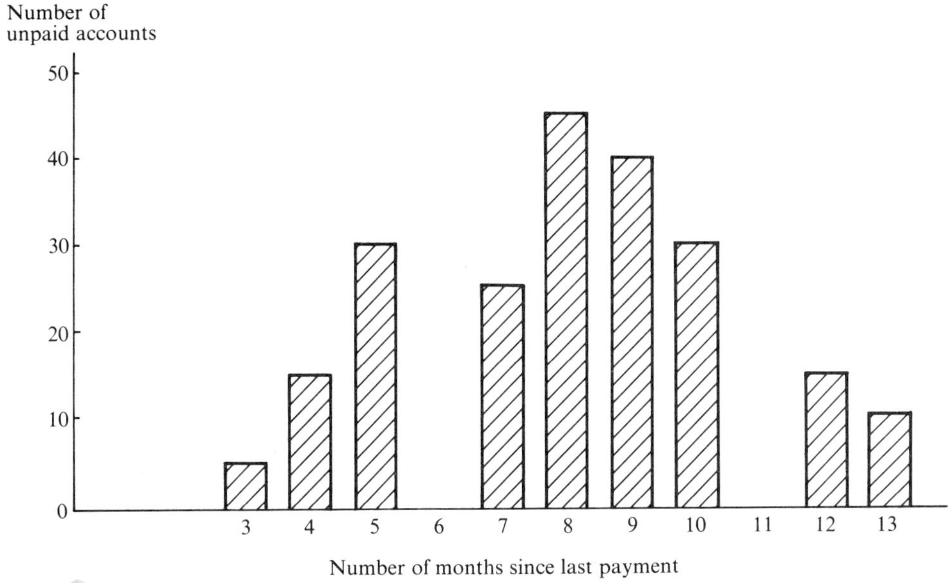

Figure 7.2

Bar Chart

In the typical situation, a simple grouping table constructed for a population is much too long and it is difficult to discern the underlying distribution of the data. It is usually better to organize the data into a more condensed frequency distribution table and exhibit the population distribution in a histogram. It is sometimes considered useful to approximate the histogram with a smooth, continuous distribution curve, as illustrated in Figure 7.3. The continuous curve is often sketched freehand. Sometimes it is fitted to the data mathematically.

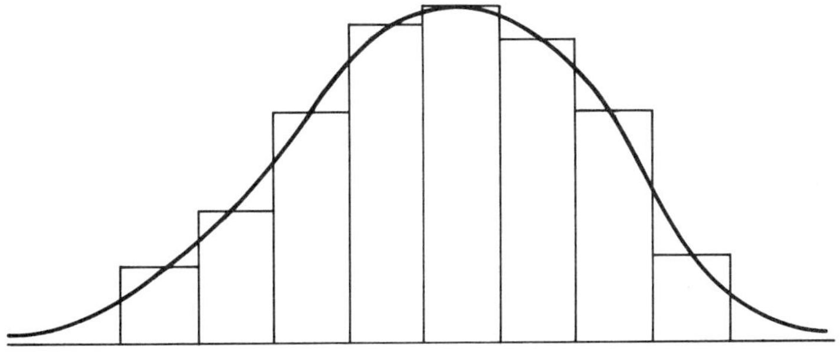

Figure 7.3

Approximating a Histogram with a Continuous Curve

Various types of distributions were presented in histogram form in Figure 7.1. Corresponding continuous distributions are shown in Figure 7.4.

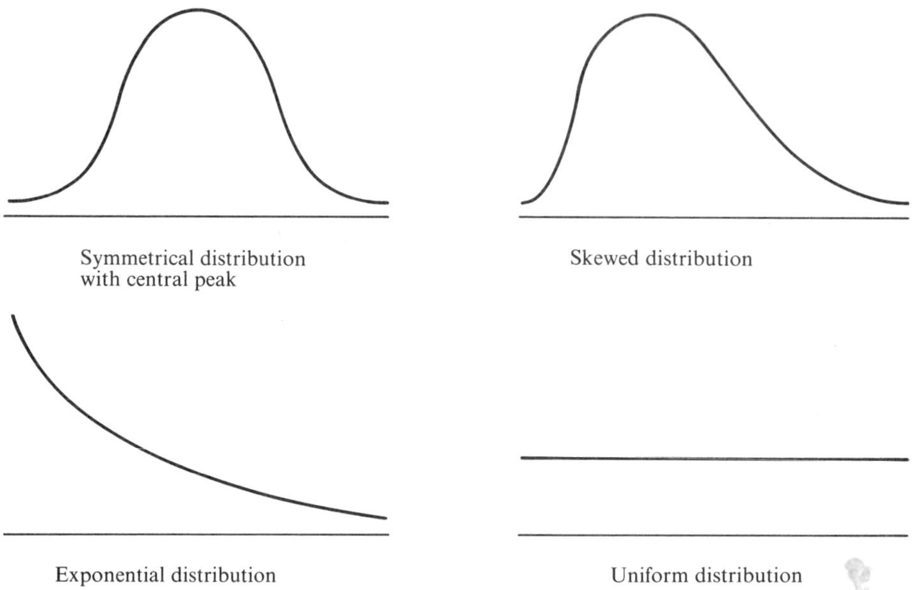

Figure 7.4

Various Continuous Distributions

In the case of sample data, the distribution as displayed by a histogram may be considered to represent the distribution of the population from which the sample was selected. However, there is always a question as to whether the observed distribution shape is characteristic of the population or merely characteristic of the particular sample selected.

Often, when dealing with a sample, a theoretical distribution of an infinite population is used to represent the population under study. Sometimes a discrete theoretical distribution is appropriate, such as the *binomial population distribution.* Other times, a continuous theoretical distribution is appropriate, such as the *normal distribution.* These theoretical distributions represent theoretical or mathematical models and are discussed in chapter 13.

7.3 Distributions of Qualitative Data

Our consideration of frequency distributions and the presentation of these distributions in graphic form dealt with quantitative (interval and ratio scale) data. How do we organize, summarize, and graph qualitative (nominal and ordinal scale) data?

Suppose it is desired to determine the distribution by color of the 83 pairs of shoes sold by a store. The scale of colors, of course, is a nominal scale. Table 7.1 presents the frequency distribution for these data. Notice that the colors are in alphabetical order (as a convenience for the user of the table). Table 7.2 presents a frequency distribution summarizing the attitudes of voters toward a new tax proposal. The attitude scale represents an ordinal scale.

Table 7.1

Frequency Distribution: Numbers of Pairs of Shoes Sold during the Last Week of April, by Color

Color	Number of Pairs Sold	Percentage Distribution
Beige	18	22
Black	30	36
Green	5	6
Red	10	12
White	20	24
Totals	83	100

Table 7.2

Frequency Distribution: Voters' Attitudes toward a New Tax Proposal

Attitude	Number of Voters	Percentage Distribution
Very much in favor	930	3.1
In favor	2,160	7.2
Neutral	3,000	10.0
Opposed	15,750	52.5
Very much opposed	8,160	27.2
Totals	30,000	100.0

Clearly, as indicated in the table, the ordering of the scale categories ("very much in favor," etc.) is largely determined by the scale itself. Another listing sequence possible is to reverse the order. A suitable graphing form for qualitative data is a bar chart as previously illustrated (Figure 7.2).

Review

Concepts and Terminology

Bar chart
Binomial population distribution
Continuous distribution
Exponential distribution
J-distribution
Negatively skewed distribution
Normal distribution
Positively skewed distribution
Rectangular distribution
Skewed distribution
Symmetrical distribution with central peak
Uniform distribution

Problems

1. Sketch a histogram for a population of prices which has the following:
   (a) symmetrical distribution with a high concentration of prices in the center
   (b) symmetrical distribution with a low concentration of prices in the center and a number of extremely low and also extremely high prices
   (c) negatively skewed distribution with a high peak
   (d) J-distribution
   (e) uniform distribution
2. Suppose you have a list of the population of heights of ten-year-old boys. Generally speaking, what purpose would be served by applying statistical methods to these data? Give three examples.
3. What is a statistician generally interested in when he analyzes sample data?
4. "The shape of the distribution underlying a set of population data actually depends on the particular frequency distribution table constructed and the histogram constructed from this table." True or false? Discuss briefly.
5. If a bar chart is used to display a population distribution as indicated in a simple grouping table, is it necessary to reserve spaces for values with zero frequency? Why?
6. Sketch a smooth, continuous curve for a population of weights which has:
   (a) symmetrical distribution with a central peak, a low concentration of weights in the center and a number of extremely light weights, and also extremely heavy weights
   (b) negatively skewed distribution with a high peak
   (c) symmetrical distribution with a high concentration of weights in the center
   (d) rectangular distribution
   (e) exponential distribution
7. Name a theoretical distribution model appropriate for (a) discrete data, (b) continuous data.
8. If a bar chart were to be constructed for the data in Table 7.1, would it be more useful to show the number of pairs sold with the colors arranged in alphabetical order along the horizontal scale of the chart or with the colors arranged in order of the number of pairs sold for each color (say, from highest to lowest)? Why?

9. Construct a bar chart for the following distribution. (Arrange the law firms along the horizontal axis to provide the most meaningful chart.)

Law Firm	*Number of Cases Handled*
A	20
B	40
C	15
D	50
E	10
F	25
G	30
H	18
I	42

8

Averages

The most frequently used descriptive measure of a set of data is the *average*, often called a *measure of central tendency*. We speak of average height, average price, average grade, etc. What do we mean by "average"? Usually, we mean some quantity that is typical of a set of data or that is, in some way, located in the center of the data. Generally, by an "average" we mean a single value that in some sense summarizes a set of data.

For example, if the average age of ten machines owned by a firm is 9 years, this single figure gives us the understanding that these machines are pretty old and, perhaps, may soon need replacement. Of course, we know that some of the machines have been in service less than 9 years and some more. However, knowledge of the average age of the machines provides the impression that the machines are, generally, old.

Actually, there is more than one average which can be computed for a set of data. (Statisticians often refer to a set of data as a set of *observations*). The *arithmetic mean* is by far the most frequently used and most important average. The *median* is also an important type of average, but not nearly as important as the arithmetic mean. The *mode* is the simplest average to determine, but is of limited value. The *geometric mean* is a special-purpose average which is useful in special situations.

Each of the averages we will discuss represents the central tendency (the center) of a set of data. However, the notion of central tendency is different in each case, as will be evident from the discussions in the balance of this chapter. Although all averages may represent "typical values" of a set of data, this is not always true for any average. For example, consider the following four stock prices (in dollars): 30, 60, 150, 200. It is possible, of course, to find a central value (the central tendency) for these prices. This would be somewhere between \$60 and \$150. However, it is clearly impossible to find any single price which is typical of the set.

8.1 Arithmetic Mean

When we speak about an average, we nearly always mean the arithmetic mean. Statisticians usually refer to this average merely as the *mean*. If we wanted to compute the average for the three weights (in pounds)—5, 8, 11—, most of us would add the weights and divide the total by 3 to obtain $24 \div 3 = 8$. This is the arithmetic mean or the mean weight. It is as simple as that.

When the mean is computed for a set of population data, it represents the population mean. When the mean is computed for sample data, we call the result the sample mean. Based on certain technical criteria, the sample mean can be shown to be the best single estimate of the mean for the population from which the sample was selected.

Suppose a sample of apples were selected from a carload and the mean diameter were found to be $2\frac{1}{2}$ inches. We may estimate that the mean diameter per apple for the entire carload (the population) is $2\frac{1}{2}$ inches. Or, assume it were found that the mean production rate for a sample of workers selected from the large work force in a firm is 44 units per hour. Then, we may estimate that the mean production rate for the entire work force is 44 units per hour.

The arithmetic mean is a most convenient type of average because of its remarkable ease of computation in certain situations. Let us consider a few.

Example 1 Suppose a shovel is manufactured with different types of handles and different types of blades. If the mean weight of the handles for a production lot is known to be 8 ounces and the mean weight of the blades is 10 ounces, we may compute the mean weight of a shovel (handle and blade) as 8 oz + 10 oz = 18 oz. It is not necessary to determine the total weight of each shovel and then compute the mean for these total weights.

Or, if we know that the mean price of hat and dress sets sold in a department store is \$85 and the mean price of a hat alone is \$10, we may immediately determine that the mean price of a dress is \$85 − \$10, or \$75. It is not necessary to subtract the hat price from the price of each set to determine the price of the dress and then compute the mean price of a dress.

Example 2 Suppose a firm produces different styles of small radios and the mean wholesale price is \$22 per set. If the price per set is raised by some constant amount, say \$5 per set, the new mean price per set is raised by the same amount. In our example the new mean price is \$22 + \$5, or \$27.

Or, if the price per set is reduced by a constant amount, say \$3, the new mean price per set is \$22 − \$3, or \$19.

Example 3 If the test scores on an accounting test for a class of students has a mean of 47 and the professor decides to double the test score for each student, the new mean test score is also doubled and becomes 47 × 2, or 94.

Suppose the mean time required to produce a product is 38 minutes per unit. If a new process reduces the time required to produce each unit by one-third, the new mean time to produce a unit is computed as 2/3 × 38 minutes, or 25.3 minutes.

Example 4 Assume a firm has three plants which produce the same product. In plant A with 30 workers, the mean production per worker for a given week was 150 units. In plant B with 20 workers, the mean production per worker for that week was 120 units. In plant C with 40 workers, the mean production per worker for that week was 260 units. What was the mean production per worker for the three plants combined? Statisticians call this mean the *general mean*, the *combined mean*, or the *overall mean*.

The computation is quite simple. If plant A produced 150 units per worker on the average with 30 workers, total production for the week for this plant was 150 × 30, or 4,500 units. Similarly, total production for plant B was 120 × 20, or 2,400 units, and for plant C, it was 260 × 40, or 10,400 units. Then, total production for the three plants together was 4,500 + 2,400 + 10,400, or 17,300 units. The total number of workers in the three plants was 30 + 20 + 40, or 90. Finally, the overall mean production rate for the firm was 17,300 units produced divided by 90 workers employed, or 192.2 units per worker.

Statisticians sometimes compute a *weighted mean* to express the appropriate average. For example, suppose a $500 investment in bonds yields a 6% return, a $1,000 bond investment yields 8%, and a third bond investment of $2,000 yields 5%. What is the yield for the total investment of $3,500? It would be a mistake to compute the *simple mean* of the three bond yields (6% + 8% + 5% = 19; then, 19 ÷ 3 = 6.33%). This mean yield (6.33%) would represent the mean for the total investment only if the amount invested for each yield were the same. However, in our illustration, the amount of investment at each yield is different. The $2,000 invested for a 5% yield gives more weight (importance) to 5% than the $500 invested for a 6% yield gives to 6%, etc. Therefore, we must take these *weights* into consideration in determining the mean yield for the three investments.

This is accomplished by, first, multiplying each yield by its weight (amount invested). Writing 6% as 0.06, 8% as 0.08, and 5% as 0.05, we compute $500 × 0.06 = $30, $1000 × 0.08 = $80, and $2,000 × 0.05 = $100. Then, $30 + $80 + $100 = $210 represents the total amount earned for the $3,500 invested. Finally, compute the mean yield for the total investment as $210 ÷ $3,500 = 0.06, or 6%. This is the *weighted mean yield* because each yield is weighted by a measure of its importance.

A slightly different use of the weighted mean is made in computing total test scores. For example, say a trade school gives an aptitude test to its seniors and the test has three parts: mathematics, finger dexterity, and spatial perception. The importance of each part is indicated by the weights assigned: 3 (mathematics), 4 (finger dexterity), and 3 (spatial perception). If a senior obtains the scores 80 (mathematics), 70 (finger dexterity), and 90 (spatial perception), his total test score is computed as the weighted mean of his scores for the three parts.

The total test score is computed as follows. First, multiply each score by its weight to obtain $80 \times 3 = 240$, $70 \times 4 = 280$, $90 \times 3 = 270$. Then, $240 + 280 + 270 = 790$ and the weights add to 10 $(3 + 4 + 3)$. Finally, the *weighted mean* of the three part scores is $790 \div 10$, or 79. This is the senior's total test score.

8.2 Median and Other Position Measures

The *median* is a measure of position since it is centrally located in a set of observations (data). Say the number of crimes occurring in five cities last month were (per thousand population) 8, 7, 10, 3, and 9. The median number of crimes (per thousand population) is determined by ranking the five observations (3, 7, 8, 9, 10) and finding the middle one. The third item of data in the ranked sequence (8) is the middle observation. This is the median. It is very simple to find the median.

It was easy to see that for five items of ranked data (observations), the third observation is the middle one. Which is the middle observation for a set of twenty-seven observations? We may find the middle observation rapidly as follows: $27 + 1 = 28$, $28 \div 2 = 14$. The fourteenth observation in the ranked sequence is the median. In a set of, say, seven observations, the middle one is the fourth observation ($7 + 1 = 8$, $8 \div 2 = 4$).

Suppose you want to determine the median for a set of four observations. Clearly, in an even-numbered set of observations there is no middle one. Say for four observations we compute $4 + 1 = 5$, then, $5 \div 2 = 2\frac{1}{2}$. What does this mean? This indicates that the median is between the second-ranked and third-ranked observations (the two middle observations). Assume the four observations are the prices (in cents) 10, 15, 12, 7. Ranking, we have 7, 10, 12, 15. Then, the second and third observations are 10¢ and 12¢. Finally, the median is located midway between these two values and computed as $10 + 12 = 22$, and $22 \div 2 = 11$ (median).

It is possible to find repeated values (*ties*) in a set of observations. Say we have the six weights (in pounds) 7, 2, 8, 2, 7, 6. How is the median determined? Ranking, we obtain 2, 2, 6, 7, 7, 8. Notice that the tied observations are placed next to each other in the ranking. Then, $6 + 1 = 7$ and $7 \div 2 = 3\frac{1}{2}$; so, the median is between the third and fourth observations (6 and 7). Finally, the median is computed as $6 + 7 = 13$, $13 \div 2 = 6.5$ (median).

The median is at the half-way mark in a ranked set of observations. The median is also called the *50th percentile*. Generally, the 50th percentile is defined as that value where 50% of the observations in a set have a lower value. This is called a *position measure* or a *percentile*.

There are other position measures which are sometimes used. The 6th percentile is that value where 6% of the observations have a lower value, etc. Percentiles that are frequently used in the social sciences (education, sociology, psychology) are the 25th, 50th, and 75th percentiles.

The 25th percentile is also called the *first*, or *lower quartile*. It is that value where one-fourth of the observations have a lower value. The 50th percentile, or the median, is also called the *second quartile.* The 75th percentile is called the *third,* or *upper quartile.* It is that value where three-quarters of the observations have lower values (or, looking at it the other way, one-quarter of the observations have higher values). The quartiles divide a set of observations into four equal parts.

Other frequently used sets of percentiles are the 10th, 20th, 30th, etc., percentiles. These are also called *deciles.* The 10th percentile is called the *first decile,* the 20th percentile is called the *second decile,* and the 90th percentile is called the *ninth decile.* These percentiles (or deciles) are interpreted in the same way as the others. The deciles divide a set of observations into ten equal parts.

8.3 Mode

The *mode* is probably the easiest average to understand and the easiest to determine. The mode is the value in a set of observations that appears most often. In the set or prices (dollars)—12, 8, 7, 12, 14, 12, 15—$12 is the mode (or *modal price*) because it appears three times, whereas the other prices appear less often.

Sometimes a set of observations does not have a mode. For example, in the set of weights (pounds) 16, 18, 16, 17, 18, 17 there is no mode because each weight appears an equal number of times (twice). Sometimes a set of observations has more than one mode. The set may be *bimodal* (two modes) or *multimodal* (many modes). In the set of prices (cents) made up of two 3s, six 5s, one 7, one 10, one 12, seven 14s, one 15, and one 18, there are two modes. One is 5¢ for the price range 3¢ to 7¢. The other is 14¢ for the price range 10¢ to 18¢.

A multimodal set of observations often indicates that the data are *heterogeneous* (not *homogeneous*). For example, if a set of stock prices has two modes, there may be a mixture of common and preferred stock prices and one modal price refers to the common stocks, the other to the preferred stocks. Or, the weights for college seniors in a university may be bimodal where one mode refers to the weights of college men and the other to college women.

8.4 Geometric Mean

The *geometric mean* is a specialized type of average. It is sometimes used in index number construction. However, it is used more often to compute the *average rate of change* or the *average rate of growth.* It is also used to compute an *average ratio.*

The geometric mean is more difficult to compute than the other averages discussed. However, we can show the computation for some simple illustrations. For example, for two prices, 2¢ and 8¢, the geometric mean is computed by first obtaining the product $2 \times 8 = 16$ and then the square root of the product which is 4; so, 4¢ is the geometric mean price. (The square root of 16 is 4 because $4 \times 4 = 16$; the square root of 9 is 3 because $3 \times 3 = 9$; etc.) The geometric mean for three weights (in pounds), 4, 1, and 2, is computed by first obtaining the product $4 \times 1 \times 2 = 8$ and then the cube root of the product which is 2; so, 2 pounds is the geometric mean of the weights. (The cube root of 8 is 2 because $2 \times 2 \times 2 = 8$; the cube root of 27 is 3 because $3 \times 3 \times 3 = 27$, etc.)

Suppose a city has a population of 2,000 in a given year, say year 1, then grows 10% the next year (year 2) to a population of 2,200. In the third year, year 3, it grows 15% to 2,530 and in year 4 an additional 60% to 4,048. What is the average yearly rate of population growth for this city during these four years? The geometric mean of *link relatives* is the appropriate measure to compute for this purpose. A link relative is computed as the amount for a time period divided by the amount for the previous time period. The link relative for year 2 is 2,200 ÷ 2,000 = 1.10; for year 3, it is 2,530 ÷ 2,200 = 1.15; and for year 4, it is 4,048 ÷ 2,530 = 1.60. The link relative is not needed for year 1 for our purpose. The geometric mean of the link relatives 1.10, 1.15, and 1.60 provides the basis for measuring the average yearly rate of population growth.

Computation of the geometric mean generally involves the use of logarithms and is somewhat complicated and laborious. It is best to leave this to the statistician. (Refer to **Note 8A** at the end of this chapter for computation of the geometric mean.)

The geometric mean of the link relatives in our illustration comes out to be 1.265, which indicates that on the average, during the four-year period, the population size for a year was 1.265 (or 126.5%) of the population size the year before. In other words, the average yearly rate of population growth was 0.265, or 26.5% per year. We can test this average by increasing the year 1 population by 26.5% a year for the next three years. This results in a year 4 population of 4,049 which differs negligibly from the true figure (4,048), due to rounding.

It would be incorrect to use the arithmetic mean of the link relatives to obtain the average yearly growth rate. If computed, the arithmetic mean comes out to be 1.283 which indicates an average yearly growth of 0.283 or 28.3%. If the year 1 population is increased 28.3% a year for the next three years, the resulting year 4 population comes out to be 4,224. This is clearly higher than the year 4 population of 4,048.

Suppose $15,000 were invested for nine years and grew to $35,000. What was the average yearly rate of return on this investment? The geometric mean rate of return should be computed. It comes out to be 9.9% a year.

Another application of the geometric mean is to compute the average ratio. Suppose a firm computes the ratio of capital costs to labor costs for each of ten plants operated by the firm. What is the average ratio of these costs for the ten plants? The arithmetic mean of the ten plant ratios is not the correct average to use. The geometric mean is the proper average for ratios.

8.5 Which Average Should You Use?

Clearly, the averages (mean, median, mode, geometric mean) are quite different. How do they compare? In a given problem, which is best to use?

The arithmetic mean is the *equilibrium point* (*point of balance*) of a set of observations. For example, for the set [2, 6, 13] the mean is 7. Suppose we have a weightless bar and three cylinders of equal weight placed on the bar as shown in Figure 8.1. As shown in the chart, if the cylinders are placed on the bar at positions corresponding to the observed values (2, 6, 13) and the fulcrum is placed at the location of the mean, the bar will be in perfect balance. This is a physical interpretation of the mean.

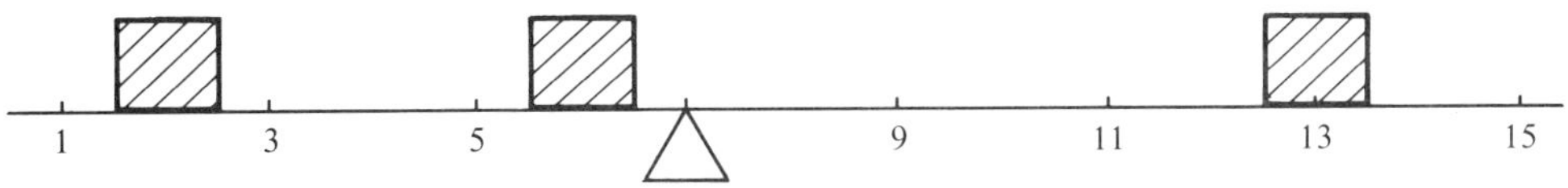

Figure 8.1

Representation of the Arithmetic Mean (7) as the Equilibrium Point of a Set of Three Quantities (2, 6, 13).

The mean is a very *sensitive average* because it is determined by each observation in a set. In the set [2, 6, 13], if any one of these values changes, the mean will also change. Say 2 is changed to 5. Then, for the new set [5, 6, 13] the mean is 8. This is an important and usually desirable property of the mean. However, if a set of observations contains *extreme values*, the mean is not the best average to use to describe (summarize) the set. For example, for the set [2, 4, 6, 500] the mean is 128, which of course is not a representative value for this set.

The median is a better descriptive measure in such instances because it is not affected by extreme values. For example, in the set of observations [2, 4, 6, 12, 18], the median is 6. If 18 is changed to 1,800, or 18,000, or even 18,000,000, the median is still 6 because it is still the middle value.

The mode is *practically insensitive* to the particular values in a set of observations because the mode is the value which appears most often. If 20 is the mode in a set of observations, it will continue to be the mode no matter how the observations change, just as long as 20 remains the most frequent value in the set.

In a symmetrical distribution with central peak, as illustrated in Figures 7.1 and 7.4 (pages 44 and 46), the mean, median, and mode are all equal. In such instances, it does not matter which average is used. The mean is, generally, the preferred average and should be used as far as possible. However, if it is desired to compute an average to describe a set of data, the median should be used if the set contains extreme values. The mean is usually unsatisfactory in such situations.

Actually, the mean, median, or mode determined from sample data may be used as an estimate of the population mean under certain conditions. However, for technical reasons, the sample mean is nearly always used for this purpose because it provides the best type of estimate. This is a most important advantage of the mean and is the reason for its widespread use.

Only in certain special problems is the mode the proper average to use. These are problems where interest centers on the greatest frequency. For example, say a retail outlet determines the number of customers returning merchandise each hour of the day for a period of time and finds that the largest number of returns occurs between 2 P.M. and 3 P.M. (the modal hour). Then, if he wishes to reduce his labor costs for merchandise returns, the owner could permit returns only between 2 P.M. and 3 P.M. daily and satisfactorily serve the greatest number of his customers.

The mean can be determined only for ratio and interval scale data (quantitative data). The median can be determined for these data and also for ordinal scale data. Table 8.1 presents the quantity of a food item sold in a food store, by grade. The scale of grades in this table is an ordinal scale, with "extra fancy" the highest grade and "utility" the lowest. The cumulative quantity column in the table shows the accumulated quantity sold as you go down the grade scale (80 lb "extra fancy," 80 + 120 = 200 lb "fancy" or better, etc.). Since a total of 370 lb was sold, the position (location) of the median grade is found by computing 370 + 1 = 371, 371 ÷ 2 = $185\frac{1}{2}$. Then, looking down the cumulative quantity column, notice that 200 lb were sold at the "fancy" grade or better, including the 185th and 186th pounds. Therefore, "fancy" is the median grade sold. Notice in the table, "fancy" is also the modal grade sold. Clearly, it is not possible to compute the mean for this problem.

Table 8.1

Quantity of a Food Sold, by Grade

Grade	Quantity Sold (Pounds)	Cumulative Quantity
Extra fancy	80	80
Fancy	120	200
Quality	100	300
Utility	70	370

Only the mode can be determined for nominal scale data (classification data). Say a luggage store sells, during a week, 3 white cosmetic cases, 6 green ones, 20 red ones, and 4 blue ones. The colors make up a nominal scale and the only "average color" which can be determined is the modal color, which in this example, is red.

Usually, you do not have a choice when the geometric mean is the proper average to compute. The only average appropriate for rates of change or rates of growth and for ratios is the geometric mean.

Review

Concepts and Terminology

Arithmetic mean
Average
Average rate of change
Average rate of growth
Bimodal
Combined mean
Decile
General mean
Geometric mean
Link relative
Mode
Multimodal
Observation
Overall mean
Percentile
Population mean
Position measures
Quartile
Sample mean
Simple mean

Concepts and Terminology (*cont.*)

Mean
Measure of central tendency
Median
Weights
Weighted mean

Problems

1. Generally speaking, what is an "average"?
2. Indicate whether the following statements are true or false and defend your choice.
   (a) An average always represents a typical value for a set of data.
   (b) The arithmetic mean is computed in the same way for population data and sample data and is interpreted in the same way for both.
   (c) If a product is made of wood parts and metal parts and the only information available is the mean weight of the wood parts per product and the mean weight of the metal parts per product, we do not have sufficient information to obtain the mean weight per product.
   (d) If we compute the general mean family income for cities A, B, and C based on the mean family income for each city, this is the same as the mean computed directly for the families in these cities.
   (e) The median is just as sensitive to extreme values in a set of data as the arithmetic mean.
   (f) The median can be computed for a set of data even if there are repeated values (ties).
   (g) The mode is always centrally located in a set of data.
   (h) Three-fourths of a set of values are higher than the lower quartile.
   (i) A multimodal set of values may indicate that the set is not homogeneous.
3. A class of students obtained the following grades on a short quiz: 10, 8, 10, 7, 5. Find the arithmetic mean, median, and mode.
4. In Problem 3, which average or averages would you say are typical of the set of grades?
5. Find the mean, median, and mode for the population of prices (in dollars): 20, 10, 20, 12, 10, 12.
6. The following sample of lengths (inches) was selected from a population: 5, 4, 5, 8, 2, 5, 32, 2, 9. Find the arithmetic mean, median, and mode.
7. In Problem 6, would you say that the arithmetic mean represents a typical length for the sample? Why?
8. A sample of 10 accounts showed a mean payment of $200 last month. What would you estimate as the mean payment for the population of accounts from which the sample was selected?
9. If the mean time required to assemble a product is 48 minutes and the mean time required to assemble one part of the product is 21 minutes, what is the mean time required to assemble the balance of the product?
10. The mean weight of a metal toy is 10 ounces.
    (a) If a heavier metal is used and it is found that the new metal is 50% heavier, what is the mean weight of the toy?

(b) If a new 3 ounce part is added to the toy and the original metal is used, what is the mean weight of the toy?

11. If $100 is spent on 9 items, what is the mean price per item?

12. If the mean weight of 40 items is 12 pounds, what is the total weight of the items?

13. Class A in French has 10 students and obtained a mean test grade of 40. Class B in French has 5 students and obtained a mean grade of 60 on the same test. Class C, with 15 students, obtained a mean grade of 70 on the same test. What is the mean grade for the three classes combined?

14. What do statisticians call the mean obtained in Problem 13?

15. If you invested $300, $1,000, and $1,200 in three different ventures and after a year earned 4%, 8%, and 10%, respectively, what would be the average percentage earned for the $2,500 invested?

16. If you invested $500 in stock A, $500 in stock B, and $500 in stock C and obtained a return of 10%, 5%, and 6% after a year, would the average percent earned be determined by computing the arithmetic mean of the three percentages? Why?

17. A student obtained 40 on part one of a test, 70 on part two, and 90 on part three. If the weights assigned are 5 for part one, 2 for part two, and 3 for part three, what is the student's total test score?

18. What is a percentile?

19. How are quartiles and deciles related to percentiles?

20. What percentage of a population of test scores are—
(a) above the 18 percentile?
(b) below the upper quartile?
(c) above the 7th decile?
(d) between the lower and upper quartile?
(e) between the 1st and 9th decile?
(f) above the 1st quartile?

21. Could the median and mode be equal? Defend your answer.

22. Would you expect more than one mode in a heterogeneous set of data? Why?

23. Which average is appropriate to compute the average rate of increase in viewers of a certain television program over a period of time?

24. Why is the arithmetic mean a sensitive measure?

25. Is the mean or median a better average to describe the set of error counts 24, 27, 68, 15, 20? Why?

26. In a set of five weights, if the smallest weight is increased so that it becomes the second smallest, will the—
(a) mean weight change?
(b) median weight change?
(c) modal weight change?

27. If a set of data has a perfectly symmetrical distribution with a central peak, which is the best average to use to represent the data—the mean, median, or mode?

28. The attitudes of 105 club members toward a certain candidate for club president were tabulated as follows:

Attitude	*Number of Members*
Very much against	30
Against	35
Indifferent	25
In favor	10
Very much in favor	5

(a) Determine the median attitude.
(b) Determine the modal attitude.

29. Students in a trade school were classified according to trade studied as follows: plumbing 22, air conditioning 18, radio and TV repair 29, carpentry 40, bricklaying 15, auto mechanics 30. If you were asked to determine an "average" trade studied, which average would you determine? Determine this "average" and indicate what it means.

30. The diameters of a sample of 25 white blood cells of a man were measured, and they totalled 275 micrometers. What would you estimate as the mean diameter of this man's white blood cells?

31. The number of cases heard in the state supreme court last month was 2 for Maine, 4 for Arkansas, 4 for Rhode Island, 4 for Vermont, 3 for Missouri, 4 for New Mexico, 1 for Colorado, 1 for Iowa, 1 for Wyoming, and 3 for North Dakota. Find the mean, median, and modal number of cases heard by these states.

32. Twenty small willow trees had a combined weight of 100 lb. After each tree was planted in a small plot and allowed to grow for a period of time, the combined weight was 160 lb.
(a) What was the mean weight per tree before planting?
(b) What was the mean weight after the period of growth?
(c) What was the mean weight gain during the period of growth?

Note 8A

Computation of the Geometric Mean

$$G' = \frac{1}{N} \times \sum \text{Log } X_i$$

$$G = \text{antilog } G' \left(\text{or } = \sqrt[N]{X_1 \times X_2 \times X_3 \times \cdots \times X_N}\right)$$

where:

G = geometric mean
X_i = $X_1, X_2, \ldots$ (item values)
$\sum \text{Log } X_i$ = sum of the logarithms of the item values
N = number of items

Suppose we wish to compute the geometric mean of the link relatives computed in chapter 8 (1.10, 1.15, 1.60). In order to apply the foregoing equation, determine the corresponding logarithms which are 0.04139, 0.06070, 0.20412. Applying the equation, we have

$$G' = \frac{1}{3} \times \left(0.04139 + 0.06070 + 0.20412\right)$$

$$= \frac{1}{3} \times \left(0.30621\right) = 0.10207$$

Then, $G = 1.265$, as shown in chapter 8.

When the geometric mean is to be computed as a measure of the average rate of change over a period of time, it is more convenient to compute

$$G' = \frac{1}{N - 1} \times \text{Log}\left(\frac{X_N}{X_1}\right)$$

$$G = \text{antilog } G'$$

where: N = number of observations in the time period
X_1 = first observation in the time period
X_N = last observation in the time period

Let us apply this equation to obtain the average yearly rate of growth in the city's population for the four-year period discussed in chapter 8. The population for the four years was given as 2,000, 2,200, 2,530, and 4,048. So, $N = 4$, $X_1 = 2{,}000$, $X_4 = 4{,}048$. Applying the foregoing equation, we have

$$G' = \frac{1}{4 - 1} \times \text{Log}\left(\frac{4{,}048}{2{,}000}\right) = 0.10207$$

Then, $G = 1.265$, as before.

Suppose the number of orders received during the first year of operation of a mail order firm were 2,000 compared with 10,000 orders during the tenth year of operation. We may compute the average yearly rate of growth in mail orders for this firm by applying the foregoing equation with $N = 10$, $X_1 = 2{,}000$, $X_{10} = 10{,}000$ to obtain

$$G' = \frac{1}{10 - 1} \times \text{Log}\left(\frac{10{,}000}{2{,}000}\right) = 0.07767$$

Then, $G = \text{antilog } 0.07767 = 1.20$ indicates an average yearly rate of growth in mail orders of 20% a year.

9

Variation and Analysis

Data and variation usually go together. The diameter of bolts produced by a machine may be expected to vary from bolt to bolt. The number of students enrolled in a university may be expected to vary from year to year. Variation is the most important characteristic of a distribution (or a set of observations), from the point of view of statistical methods.

If the 10,000 family incomes in a city were all identically the same (say, $8500), there would be no income variation at all and no need for the use of statistical methods to analyze the city's income distribution. We would know at once all there is to know about these 10,000 family incomes. However, typically, variation does exist in a set of observations and statistical methods are required to describe and analyze the data.

Determining the amount of variation in a set of observations is important in describing the set. For example, in both set A [11, 12, 13] and set B [2, 4, 30] the mean is 12. However, set A has very little variation, and variation in set B is considerable. Therefore, knowledge that the mean is 12 provides a better description of set A than of set B.

The extent of variation in a set of observations is of major importance in sampling problems. The greater the variation, the more time-consuming and the more costly it is to adequately describe a population based on sample data, because a larger sample is required.

The simplest measure of variation is the *range* (the difference between the highest and lowest values in a set). In the set of prices (dollars) [10, 14, 25] the range is \$25 − \$10 = \$15. Sometimes, the range is expressed as the *interval* (range of values) \$10–\$25.

Certainly, if we know that the mean price in this set is \$16.33 and the range is \$10–\$25 (or \$15), we have a better description of the set of prices than if we know only the mean. However, the range is of limited value since it depends only on the two extreme values (highest and lowest) in a set.

9.1 Standard Deviation

By far the most important and most frequently used measure of variation in statistics is the *standard deviation*. Let us examine this important statistical measure.

Consider the set of weights (pounds) [1, 6, 4, 9]. The mean is 5 pounds. The weight 1 pound is 4 pounds below the mean; the weight 6 pounds is 1 pound above the mean; the weight 4 pounds is 1 pound below the mean; and the weight 9 pounds is 4 pounds above the mean. How far an observation is above or below the mean is called a *deviation from the mean*. Then, the deviations from the mean [4, 1, 1, 4] represent the variation of the set of weights around their mean. (Note: Statisticians write these deviations as $-4, 1, -1, 4$. The negative quantities indicate deviations below the mean and the positive quantities indicate deviations above the mean.)

The standard deviation is a single quantity which, *in a special way*, averages the deviations around the mean for a set of observations. For our set of weights, the standard deviation is 2.9 pounds, which is a sort of average amount by which a weight deviates from the mean. Clearly, then, the standard deviation computed for a set of observations is an average measure of how far an observation is from the mean (above or below).

If the standard deviation for a set of observations is zero, it means that the observations do not deviate at all from the mean. In other words, all the observations are equal to the mean and to one another. There is, therefore, no variation at all in the set (and no need for the use of statistical methods to describe the set of data). The smaller the standard deviation, the less is the variation of the observations around its mean. The standard deviation is never a negative quantity. It should be noted that the standard deviation is the most important measure in all of statistics.

The computation of the standard deviation is somewhat complicated. (Computation of standard deviation is covered in **Note 9A** at the end of the chapter.)

The standard deviation computed for a population is a measure of the extent of variation of the quantities in the population. The standard deviation computed for a sample can be used as an estimate of the standard deviation *in the population from which the sample was selected.*

Suppose a sample of 75 men attending night school in New York City were asked how many times they were absent from school during the previous three months. Based on the replies, it was determined that the mean number of absences was 12 and the standard deviation was 4. Using these sample results, we may estimate that for the entire population of male night school students, the mean number of absences is 12 and the standard deviation is 4. This indicates, on a kind of average basis, that the number of absences of a male night school student is 4 more or 4 less than the estimated mean of 12 absences (a deviation of 4).

Statisticians often speak of the *variance*, which is the square of the standard deviation (the standard deviation multiplied by itself). For example, if the standard deviation for a set of observations is 20, then the variance is $20 \times 20 = 400$. However, for our purposes, the standard deviation is an easier concept to understand.

In certain situations, the standard deviation is easy to determine. Let us consider a few of these.

Example 1 Say the standard deviation for a set of test scores is 25. If the teacher decides to raise each score by some constant amount, say 10 points, in order to raise the general level of class scores, the standard deviation does not change. Or, if the class is penalized by reducing each score by some constant amount, say 5 points, the standard deviation also remains the same.

Example 2 Suppose the standard deviation for the common stock prices in an investment portfolio is \$50. If, as a result of sharp price advances, each price is multiplied by some constant, say 3, the standard deviation is also multiplied by the same constant. In our illustration, the new standard deviation is \$50 × 3 = \$150. Or, if all the stock prices show the same percentage decline (say 15%), the standard deviation will also be 15% smaller.

9.2 Other Measures of Variation

The standard deviation is typically used as a measure of variation in analyses where the arithmetic mean is used as the average. Where the median is used as the average, other measures of variation are more appropriate, such as the *interquartile range*, the *quartile deviation*, or the *interdecile range*. These measures are often used in social sciences such as education, psychology, and sociology.

The *interquartile range* is computed as the difference between the upper and lower quartiles of a set of data. Suppose that for a set of grades in a business forecasting course, the lower quartile is 55 and the upper quartile is 80. Then, the interquartile range for these grades is $80 - 55 = 25$, indicating that the middle 50% of the grades has a spread of 25 points. We also know from this information that the middle 50% of the grades are between 55 and 80.

The *quartile deviation* is computed as half the interquartile range. In our illustration, the quartile deviation is $25 \div 2 = 12.5$. The quartile deviation is sometimes called the *semi-interquartile range*.

The *interdecile range* is computed as the difference between the ninth and first deciles. For example, if the first decile for the scores on a clerical aptitude test is 30 and the ninth decile is 95, the interdecile range is computed as $95 - 30 = 65$. This indicates that the middle 80% of the clerical aptitude test scores has a spread of 65 points. We also know from this information that the middle 80% of the scores are between 30 and 95.

9.3 How to Evaluate Variation

The different measures of variation we have just discussed are very useful statistical measures that offer considerable help in analyzing a set of data and its distribution.

However, taken by themselves, they may be difficult to evaluate. For example, suppose a firm manufactures washing machines and it is determined that the standard deviation of the average monthly prices of washing machines sold by the firm during the last three years is $50. Does this indicate considerable variation in average monthly prices or very little variation?

Suppose this firm also produces dryers and for the same three-year period average monthly prices for dryers sold have a standard deviation of $30. Should we conclude that there was more variation among average monthly prices for washing machines than for dryers?

Let us consider the washing machine prices first. If the mean price of the machines sold during the three-year period were $100, a standard deviation of $50 would appear to indicate much more variation in washing machine prices than if the mean price were $500. And, if the mean price were $50,000, a standard deviation of $50 would indicate hardly any variation in average monthly prices of the washing machines. The standard deviation computed for a set of data can be evaluated only when the average of the quantities involved in the data is taken into consideration. The problem of evaluating standard deviations becomes more difficult when two or more sets of data are compared.

The statistician deals with such problems by expressing the measure of variation as a percentage of the average. That is, he computes the *relative variation* for the set of data. The most important measure of relative variation is the *coefficient of variation.* This measure is very simple to compute.

In our washing machine illustration, the standard deviation is $50. Let us assume the mean price were $100. Then, the coefficient of variation is computed as $50 \div 100 = 0.5$ and $0.5 \times 100 = 50\%$ (the coefficient of variation). This indicates that the standard deviation is 50% of the mean. If the mean price were $500, the coefficient of variation would be computed as $50 \div 500 = 0.10$ and $0.10 \times 100 = 10\%$. So, the standard deviation is only 10% of the mean in this case. If the mean price were $50,000, the coefficient of variation would be computed as $50 \div 50{,}000 = 0.001$ and $0.001 \times 100 = 0.1\%$, indicating that the standard deviation is only one-tenth of one percent of the mean price (hardly any variation).

Let us suppose that the mean price were $100 for the washing machines; so the coefficient of variation is 50% (as just computed), and the mean price is $50 for the dryers. Then, the coefficient of variation for the dryers is 30 (standard deviation) $\div$ $50 = 0.6$ and $0.6 \times 100 = 60\%$. Therefore, average monthly dryer prices fluctuated (varied) relatively more (60% of the mean) than average monthly prices for washers (50% of the mean).

If the quartile deviation is the measure of variation used and the median is the average, the relative variation computed is called the *coefficient of quartile deviation.* For example, if the mechanical aptitude test scores obtained by the seniors in a trade school have a lower quartile of 101 and an upper quartile of 142, we compute this measure of relative variation as follows: $142 - 101 = 41$, $142 + 101 = 243$, $41 \div 243 = 0.169$, and $0.169 \times 100 = 16.9\%$ (coefficient of quartile deviation). This indicates that the quartile deviation is 16.9% of the median for the mechanical aptitude test scores if the score distribution is symmetrical (because $41 \div 2$ is the quartile deviation, previously discussed in this chapter, and $243 \div 2$ is the median if the score distribution is symmetrical). On the other hand, if spatial perception scores for these seniors have a lower quartile of 40 and an upper quartile of 90, we compute $90 - 40 =$

50, 90 + 40 = 130, 50 ÷ 130 = 0.384, and 0.384 × 100 = 38.4% (coefficient of quartile deviation). Clearly, mechanical aptitude scores for these seniors show considerably less relative variation than spatial perception scores.

9.4 Evaluating a Single Observation

Suppose store A sold 75 TV sets during a given week. Is this a good week's business or a poor week's business? Suppose store B sold 100 TV sets that week. Could you conclude that the week in question was a better week for store B than for store A?

An observation by itself is usually not meaningful. It must be evaluated on the basis of an appropriate average and an appropriate measure of variation. Clearly, if the mean number of TV sets sold by store A weekly is 60 and the standard deviation of sets sold weekly is 5, then 75 sets sold during the week represents a very good business week. Why? Notice that 75 − 60 = 15 sets above the mean, and 15 ÷ 5 = 3 standard deviations above the mean, which indicates that 75 sets sold exceeds the mean by 3 times as much as the average deviation from the mean.

On the other hand, say the mean number of TV sets sold weekly by store B is 80 and the standard deviation of sets sold is 10. Then, 100 − 80 = 20 sets above the mean, and 20 ÷ 10 = 2 standard deviations above the mean, which indicates that 100 sets sold exceeds the mean by 2 times as much as the average deviation in weekly sales from the mean. Considering the weekly numbers of TV sets usually sold by each store and the variation usually experienced in weekly sales, store A had a better week than store B.

In the foregoing illustration, statisticians refer to 75 for store A as the week's sales in *original units* (*number of TV sets*) and to 3 as the week's sales in *standard units* (*number of standard deviations from the mean*). In the case of store B, the week's sales were 100 in original units and 2 in standard units. Converting an observation from original units to standard units provides a useful basis for comparing observations.

Suppose Fred obtained a score of 70 on a history test and a score of 70 on an economics test. Would you conclude that Fred is just as good a student in history as in economics? If history test scores had a mean of 80 and a standard deviation of 4, we may compute 70 − 80 = −10 (10 points below the mean) and −10 ÷ 4 = −2.5 (2.5 standard deviations below the mean). If economics test scores had a mean of 85 and a standard deviation of 3, we may compute 70 − 85 = −15 (15 points below the mean) and −15 ÷ 3 = −5.0 (5.0 standard deviations below the mean). After converting to standard units, it is evident that Fred did not do too well in either test and far worse in economics.

(In education and psychology it is often required to convert test scores to standard units. The method is presented in **Note 9B** at the end of the chapter.)

9.5 Evaluating the Distribution of a Set of Data

Say you know that the average (mean) number of customers entering a department store between 3 P.M. and 4 P.M. on a Tuesday is 800, with a standard deviation of 50 customers. The mean customer traffic (800) lets you know that customer traffic is quite heavy. The standard deviation of 50 indicates a coëfficient of variation of about

6%, so the variation in customer traffic from Tuesday to Tuesday from 3 P.M. to 4 P.M. is usually quite small. What else can we say about the distribution of customer traffic between 3 P.M. and 4 P.M. on a Tuesday based on the given information? Table 9.1 is useful for this purpose.

Part A of this table can be used for nearly all distributions. We need not bother with the rare exceptions. Notice that the first column lists 1.41 standard deviations first. Referring to our department store illustration, compute 50 (standard deviation) × 1.41 = 70.50. Then, 800 (mean number of customers) + 70.50 = 870.50, and 800 − 70.50 = 729.50. As a matter of convenience, replace 870.50 and 729.50 by the round figures 870 and 730. Then, 730 is the lower limit and 870 is the upper limit. These are the limits referred to in the second column of Table 9.1. We learn from Part A (second column, first row) that *at least* 50% of the distribution is contained between these limits. This means that in at least 50% of the Tuesdays between 3 P.M. and 4 P.M. between 730 and 870 customers enter the store.

Table 9.1

Estimating the Distribution of a Set of Data

Part A: All Distributions*	
Number of Standard Deviations	*Minimum Percentage* of Observations Between the Limits †
1.41	50
2.00	75
3.16	90

Part B: Normal Distributions ‡	
Number of Standard Deviations	*Percentage* of Observations Between the Limits §
.67	50
1.00	68
1.64	90
2.00‖	95
3.00	100

*Based on Chebyshev's Inequality (**Note 9C**).
†See text discussion.
‡See pages 105–107
§See text discussion.
‖Statisticians often use 1.96 standard deviations instead of 2.00, but 2.00 is adequate.

Do not overlook the *at least*, since the percentages shown in Part A are very conservative and the true percentage included between the limits could be much higher.

However, if no other information is available, the minimum percentage provides useful information indeed.

If the standard deviation in our illustration is multiplied by 3.16, we obtain 50 × 3.16 = 158. Then, the limits are 800 − 158 = 642, and 800 + 158 = 958. Using the round figures 640 and 960, we know from Part A (Table 9.1) that in *at least* 90% of the Tuesdays between 640 and 960 customers enter the store. (**Note 9C** at the end of this chapter presents the mathematical basis for the minimum percentages in Table 9.1 (Part A) and indicates how such percentages may be determined for any desired set of limits.)

Often the data involved is approximately normal in distribution. The normal distribution is discussed in chapter 13. This is a symmetrical distribution which peaks in the center and spreads out in a tail at each end. It looks something like the symmetrical distribution shown in Figure 7.4 (page 46). Many data distributions encountered in practice can be approximated by the normal distribution. For example, the weight distribution of men in a given age category, the diameter of bolts produced by a well-adjusted machine, and the scores obtained by teenagers on a mechanical aptitude test may be approximated by the normal distribution. (You will know more about this distribution after reading chapter 13.) Of course, the statistician is better able to decide whether or not the normal distribution is a useful approximation for a set of data and this decision should be left to him.

However, if a set of data has a distribution which is approximately normal, then Part B of Table 9.1 provides considerably better estimates than Part A. Suppose the average number of divorces in March for a region is 50, with a standard deviation of 5, based on March data for a long period of years. Compute 5 (standard deviation) × 2 = 10, 50 (mean) − 10 = 40, and 50 + 10 = 60. Then, if it is not known whether the number of divorces in March may be approximated by the normal distribution, referring to Part A, Table 9.1 we may state that at least 75% of the time the number of divorces in March will be between 40 and 60. On the other hand, if it is determined that the normal distribution is a reasonable approximation, referring to Part B we may state that 95% of the time the number of divorces in March will be between 40 and 60. In this case, we do not say *at least*.

Review

Concepts and Terminology

Coefficient of quartile deviation
Coefficient of variation
Deviation from the mean
Interdecile range
Interquartile range
Limits
Range
Relative variation
Sample standard deviation
Standard deviation
Standard units
Semi-interquartile range

Concepts and Terminology (*cont.*)

Lower limit
Normal distribution
Population standard deviation
Quartile deviation
Upper limit
Variance
Variation

Problems

1. It is determined that the 95 typists employed by a firm all have one year of typing experience. How can the use of statistical methods aid you in understanding the distribution of typing experience for this group of typists?
2. If the mean score on a business administration test is 82 for class A and also for class B, does this indicate that the distribution of test scores is the same for both classes? Why?
3. The amounts spent by 5 men and 5 women are:

Men	*Women*
$25	$15
$15	$30
$30	$35
$10	$10
$20	$25

   Find the mean and range (magnitude and interval) for the amounts spent by the men and women, separately.
4. The number of cancelled orders were recorded for five Thursdays as follows: 9, 4, 6, 10, 6.
   (a) Determine the mean number of orders cancelled.
   (b) Determine the range of the set of data (magnitude and interval).
   (c) Determine the deviations from the mean. (Show the deviations below the mean as negative quantities.)
5. It is determined that the standard deviation of a population of error counts is 15.
   (a) What does this tell you about the variation of the error counts?
   (b) Determine the variance.
6. If the crime rate is identically the same for a group of cities:
   (a) Determine the standard deviation for the set of crime rates.
   (b) Determine the variance.
7. Say we have two sets of measurements, set A and set B. If the individual measurements are closely clustered about the mean in set B and much more dispersed in set A, in which set is the standard deviation larger? Why?
8. A sample of weekly wages paid to construction workers has a mean of $110.15 and a standard deviation of $16.20. What would you estimate as the mean and standard deviation of the population of weekly wages paid to construction workers?
9. A population of prices has a mean of $110 and a standard deviation of $15. What is the new mean and standard deviation if—
   (a) each price is reduced by $2.50?
   (b) each price is reduced by 20%?

10. A group of test tubes containing potato-dextrose agar in varying amounts has a mean of 25 milliliters of agar and a standard deviation of 5 milliliters. What is the new mean and standard deviation if—
   (a) 5 milliliters of agar is added to each test tube?
   (b) the quantity of agar in each test tube is increased by 10%?

11. Analysis of a population of scores on a research aptitude test indicates that the upper quartile equals 150, the lower quartile equals 70, the first decile equals 50, the ninth decile equals 170, and the second quartile equals 110. Determine the following:
   (a) interdecile range
   (b) median
   (c) quartile deviation
   (d) range (interval) of the central 50% of the score population
   (e) range (interval) of the central 80% of the score population

12. The coefficient of variation for the birth rates in a group of cities was found to be 25%. Explain the meaning of this finding.

13. A sample of computer operators, group A, has a mean annual salary of $8,000, with a standard deviation of $200. A sample of computer operators, group B, has a mean annual salary of $9000, with a standard deviation of $300. Which sample shows more relative variation?

14. Soybean oil prices showed a mean of 60¢ a pint and a standard deviation of 10¢ a pint for a sample of 20 days. Cottonseed prices showed a mean of $75 a ton and a standard deviation of $15 a ton during the same period. Which of these commodities showed less price variation during the 20-day period?

15. A group of children have a mean weight of 60 pounds, with a standard deviation of 5 pounds. These children have a mean height of 48 inches, with a standard deviation of 6 inches. Do these children show more variation in height or in weight?

16. A group of men were tested for reaction time to an electric shock stimulus. Analysis of the reaction times showed a first quartile of 10 seconds and a third quartile of 20 seconds. A spatial perception test was administered to these men. Test scores showed a lower quartile of 80 and an upper quartile of 140. Do these men show less variation in reaction time or in test scores?

17. The number of cases heard by a western county court during a sample of 10 weekly periods had a mean of 70 and a standard deviation of 10. For an eastern county court during the same weeks, the mean was 60 and the standard deviation was 8. Which court had less relative variation in the weekly number of cases heard?

18. The ages of active county judges during the past year showed a first quartile of 50 and a third quartile of 60. In terms of years of experience, the first quartile for these judges was 10 and the third quartile 25. Is there more relative variation in judges' ages or in years of experience?

19. The period of time from acquisition of a plant virus by an insect to first transmission to a plant is called the latent period. It was found that for a certain variety of leafhopper insect, the latent period has a mean of 9 days and a standard deviation of 2 days. The latent period was found to be 2.8 days for insect A, 8.0 days for insect B, 10.0 days for insect C, and 12.6 days for insect D.
   (a) Transform the latent period for insects A, B, C, and D to standard units.
   (b) Which of these insects is most unusual in the length of its latent period? Why?

20. A batch of tomato tissue sections has a mean thickness of 5.01 micrometers, with a standard deviation of 0.05 micrometers. The last two sections in the batch have thicknesses of 4.05 and 5.10 micrometers.

(a) Transform these two thicknesses to standard units.
(b) Are both or either of these two thicknesses atypical of the other section thicknesses in the batch? Why?

21. On a certain day, stock A sold for \$36 a share and stock B sold for \$32 a share. Over a period of months stock A sold for a mean price of \$40 a share, with a standard deviation of \$8 a share; whereas stock B sold for a mean price of \$38 a share, with the same standard deviation. Based on this information alone, which stock represented a better buy on that day? Why?

22. Consider the following data:

Subject	*Jerry's Grade*	*Class Mean*	*Class Standard Deviation*
Chemistry	80	75	5
French	90	75	10
History	80	70	5

(a) Transform each of Jerry's grades to standard units.
(b) In which subject did Jerry do best? Why?

23. The number of automobile accidents during a given week in cities of 25,000 to 50,000 population had a mean of 60 and a standard deviation of 15. If the distribution of the number of accidents is not known, determine the following:
(a) the coefficient of variation (Does this indicate little variation from city to city? Explain.)
(b) the limits which include at least 50% of the cities with the number of accidents closest to the mean
(c) the limits which include at least 90% of the cities with the number of accidents closest to the mean

24. In Problem 23, if the distribution of the number of accidents per city is approximately normal, determine:
(a) the limits which include the middle 50% of the cities
(b) the limits which include the middle 95% of the cities
(c) the limits which include all the cities

25. The diameters of bolts produced by a machine are known to have a distribution which is not normal. If the mean diameter is 2 centimeters with a standard deviation of 0.03 centimeters, what percentage of the bolts are between 1.94 centimeters and 2.06 centimeters in diameter?

26. In Problem 25, if the diameters are approximately normal in distribution, what percentage of the bolts are between 1.94 centimeters and 2.06 centimeters in diameter?

Note 9A

Computation of the Standard Deviation

For a population:

$$\text{Standard deviation} = \frac{1}{N} \times \sqrt{N \times \sum X_i^2 - \left(\sum X_i\right)^2}$$

For a sample:

$$\text{Standard deviation} = \sqrt{\frac{n \times \sum X_i^2 - \left(\sum X_i\right)^2}{n \times (n - 1)}}$$

where N = population size
n = sample size
$\sum X_i$ = sum of the observations in a set
$\sum X_i^2$ = sum of the squared observations

Given a set of 4 prices (dollars per dozen) 4, 2, 5, and 6, compute the standard deviation. A convenient approach is to set up two columns as follows (X_i = price):

X_i	X_i^2
4	16
2	4
5	25
6	36
17	81

Then, if these prices represent a population of prices, compute

$$\text{Standard deviation} = \frac{1}{4} \times \sqrt{(4 \times 81) - (17)^2} = 1.48$$

If these prices represent a sample of prices, compute

$$\text{Standard deviation} = \sqrt{\frac{(4 \times 81) - (17)^2}{4 \times 3}} = 1.71$$

Therefore, the standard deviation for the population of 4 prices is \$1.48. However, if these prices are a sample selected from a population of prices, then the sample standard deviation, \$1.71, is an estimate of the population standard deviation.

Notice that the standard deviation is larger if we consider the prices as representing a sample. When computing a standard deviation for a sample, the purpose is to estimate the standard deviation in the population. If the equation for the population standard deviation were used for this purpose, statisticians have shown that the estimate obtained would generally be an underestimate. The equation for the sample standard deviation always yields a higher value to provide a better estimate of the population standard deviation.

Note 9B

Computation of Standard Scores

In education and psychology it is often desired to convert test scores (called *raw scores*) to *standard scores* (scores in standard units) with a predetermined mean and standard deviation. This may be accomplished as follows:

$$z = \frac{\sigma'}{\sigma} \times (X - \mu) + \mu'$$

where
μ = raw score mean
σ = raw score standard deviation
X = raw score
μ' = new mean
σ' = new standard deviation

Suppose a population of psychological test scores has a mean of 70 and a standard deviation of 8 and it is desired to convert these scores to standard scores with a mean of 100 and a standard deviation of 20. If 40 is one of the raw scores in the population, we may apply the foregoing equation to convert this to a standard score by substituting $\mu = 70$, $\sigma = 8$, $X = 40$, $\mu' = 100$, $\sigma' = 20$, to obtain

$$z = \frac{20}{8} \times (40 - 70) + 100 = 25$$

Note 9C

Chebyshev's Inequality

A useful application of the standard deviation was developed by the Russian mathematician Chebyshev: Let k = a constant equal to 1 or more.

$$p = 1 - \frac{1}{k^2}, \textit{ a proportion}$$

Then, Chebyshev's inequality, which applies to any set of quantities (as long as the standard deviation for the set is finite) states that *the proportion of the quantities in the set which are within k standard deviations from the mean is equal to p or more.*

Suppose the mean of a set of prices is 15¢, with a standard deviation of 3¢. If k is put equal to 2, $k^2 = 2 \times 2 = 4$, $1/k^2 = 1/4 = 0.25$, and $p = 1 - 0.25 = 0.75$. Then we may state that *at least* 75% of the prices in the set are within 2 standard deviations of the mean price (15¢). Computing ($k = 2$) × (standard deviation = 3) = 6, (15 mean) − 6 = 9, and 15 + 6 = 21, we may state that *at least* 75% of the prices in the set are between 9¢ and 21¢. Then, computing $1 - 0.75 = 0.25$, we can state that *at most* 25% of the prices in the set are either less than 9¢ or more than 21¢.

If $k = 1.5$, then $p = 0.56$, indicating that at least 56% of the prices are within 1.5 standard deviations of the mean. Computing ($k = 1.5$) × (standard deviation = 3) = 4.5, (15 mean) − 4.5 = 10.5, and 15 + 4.5 = 19.5, we may state that at least 56% of the prices in the set fall in the interval between 10.5¢ and 19.5¢. At most, 44% are outside of this interval (either less than 10.5¢ or more than 19.5¢).

Part C

THE MODERN BASIS FOR STATISTICAL THINKING

10

Probability: The Statistician's Language

10.1 The Role of Probability in Statistics

The statistician usually deals with sample data. Consequently, he is usually involved in making inferences (estimates) about a population based on a sample. His greatest concern is to minimize the errors involved in preparing estimates and in carrying out various decision-making procedures. We do not mean arithmetic or other computational errors, but the errors which arise as a result of the characteristics of the particular sample data used. These are called *sampling errors* and are not the fault of the statistician, but result from the normal risks involved in making statistical inferences based on sample data. We will say more about sampling errors later.

The statistician protects himself and the users of his inferences by saying something about *how good they are* and about *how much risk is involved* in using them. The language he uses for this involves *probability*.

10.2 The Meaning of Probability

The general notion of probability is familiar to most of us. We speak in terms of probability very often in our daily conversation. For example, the statements "I may buy a new car this summer," "the price is very likely to go up," "there is a fair chance that Joe will get the job" all involve probability.

A common element underlying the foregoing statements is a feeling of *uncertainty*. That is, the *outcome* in each situation is uncertain. In other words, we recognize *alternative possible outcomes* or alternative *events* that might occur.

Consider the statement, "I may buy a new car this summer." Clearly, "may" implies uncertainty. What are the alternative possible outcomes? These depend on what I have in mind. There may be no doubt in my mind that I will buy a car this summer, but will I buy a *new* car or a *used* car? I may know for sure that I will buy a new car. Will I buy it *this* summer, in the *fall*, or *next* summer?

Another common element underlying the foregoing statements is the *degree of certainty* (*or uncertainty*) involved. "I may buy a new car next summer" expresses a small degree of assurance (certainty) that I will do as I say. "The price is very likely to go up" expresses considerable assurance that the price will increase.

Generally speaking, probability deals with the degree of certainty (or uncertainty) that a particular outcome or event will occur. How is probability determined and how is it expressed? Let us consider this.

Typically, the statistician thinks in terms of *repeatable situations*. What is the probability that a well-adjusted machine will produce a defective bolt? The statistician would examine, say, the last 100 bolts produced by the machine and would determine how many are defective. The 100 bolts represent 100 repetitions of "a bolt produced by the machine." Assume 8 are defective. The proportion defective is computed as $8 \div 100 = 0.08$. Then, 0.08 is an estimate of the probability that a bolt produced by the machine will be defective. Compute $0.08 \times 100 = 8\%$. We may state that a probability of 0.08 means that 8% of the time we can expect the machine to produce a defective bolt, if the machine remains well adjusted. The larger the number of bolts examined, the better is the estimate of the probability.

What is the probability Tom will come late to work tomorrow? Say Tom came late to work 6 times during the last 150 working days. The probability we want is estimated as $6 \div 150 = 0.04$. Then, $0.04 \times 100 = 4\%$ indicates that we may expect Tom to come late to work 4% of the time.

In the foregoing illustrations, we may also state that the probability 0.08 indicates that the *chances* are 8 in 100 (8%) that the next bolt produced by the machine will be defective. Similarly, the probability 0.04 indicates that the chances are 4 in 100 (4%) that Tom will be late on the next working day.

In the same way, if a university determines that 14% of the entering freshmen drop out before reaching the senior year (based on the experience of the university for a period of years), the probability is 0.14 that a freshman entering the university will drop out before entering the senior year. In other words, the chances are 14 in 100 that he will drop out.

Notice that probability was determined on the basis of *repeated experiences* in each of the foregoing examples. In the case of Tom, we had 150 repeated experiences of "Tom coming to work," where the alternative possible outcomes were "late to work," and "not late to work." In the case of the entering freshmen in a university, the entering freshmen during the period studied were the repeated experiences and the alternative possible outcomes were "drop out before the senior year," and "do not drop out before the senior year."

If a well-balanced coin is tossed, the alternative possible outcomes are head and tail. If the coin is tossed only once, head and tail cannot both occur. Statisticians call such possible outcomes *mutually exclusive* events or possible outcomes. When Tom comes to work, the possibilities "late" and "not late" cannot both occur on a given working day; therefore, these are also mutually exclusive possible outcomes. On the other hand, if an investor watches the price of a stock listed on the New York Stock Exchange and is interested in the alternative possible outcomes "price does not go down," "price does not go up," "price does not change," these three alternative possibilities are not mutually exclusive. If the price does not change, all three possibilities occur.

Notice that when a well-balanced coin is tossed, it is just as likely that it will fall head as tail. Then, it is just as probable that the outcome will be head as that it will be tail. Statisticians call possible outcomes which are equally likely to occur *equiprobable* events or possible outcomes. On the other hand, when Tom comes to work, the possibilities "late" or "not late" are not equally likely to occur (he is more likely to come early) and, hence, are not equiprobable possible outcomes.

When the alternative possible outcomes are both mutually exclusive and equiprobable, we can determine probabilities without considering repetitions. For example, when a coin is tossed two outcomes are possible and one of these possibilities is that it will fall heads. Then, the probability of obtaining heads is computed as $1 \div 2 = 1/2$, or 0.50. If a machine is just as likely to produce a bolt which is too long, too short, or just right (three possible outcomes), then the probability that the next bolt produced is not "just right" is computed as $2 \div 3 = 2/3$, or 0.67. On the other hand, it is not just as likely that Tom will be "late" or "not late," so the probability that Tom will be late the next working day must be determined based on his past performance (repeated experiences).

10.3 Objective and Subjective Probabilities

The computation of probabilities as just discussed forms the basic method used by statisticians. These probabilities are *objective probabilities* because the *personal judgements* of the statistician are not involved. There is a growing trend for statisticians to use *subjective probabilities*. These are probabilities determined on the basis of personal judgements.

For example, what is the probability that I will buy a new car next summer? Clearly, we cannot conceive of "repeated experiences" in this case. Even if I have faced this problem many times before, I cannot use the previous experiences as a basis because each year my personal situation is different (income is different, family problems are different, car prices are different, etc.). I must use my personal judgement to estimate this probability. For example, I may feel quite sure of buying a new car next summer and may estimate the probability as 0.90 (90 chances in 100). Nevertheless, most probabilities computed by statsticians are objective probabilities.

This chapter was intended to give the nonstatistician a fairly enriched background in the basic understanding of probability as used in statistics. Of course, the statistician nearly always must deal with more complications in computing a probability than we have considered. However, we now have a good basis for understanding the meaning of the probabilities he computes. (**Note 10A** at the end of the chapter presents additional probability concepts and certain rules for computing probabilities.)

Review

Concepts and Terminology

Alternative possible outcomes
Chances
Certainty
Degree of certainty
Degree of uncertainty
Equiprobable
Event
Mutually exclusive
Objective probability
Probability
Sampling errors
Subjective probability
Uncertainty

Problems

1. The personnel office of a firm has a card prepared for each employee showing name, date of birth, job category (1, 2, or 3), and salary category (A, B, C, or D). The cards are white for female employees and green for male employees. If one card is selected arbitrarily, what are the possible alternative outcomes relating to the employee whose card is selected if you are interested in the—
(a) salary category of the employee?
(b) month of birth of the employee?
(c) job classification of the employee?
(d) sex and salary category of the employee?

2. Referring to Problem 1, the employee cards are well shuffled and one card is selected, with eyes closed.
(a) Are the possible outcomes "salary category B" and "salary category D" mutually exclusive? Why?
(b) Are the possible outcomes "month of birth is June" and "month of birth is February" mutually exclusive? Why?
(c) Are the possible outcomes "employee is male" and "job category is 2" mutually exclusive? Why?
(d) Are the possible outcomes "job category 3" and "salary category A" mutually exclusive? Why?

3. The probability that a man will slip and fall when walking to work on a rainy day is 0.30. Explain the meaning of this probability.

4. In general terms, explain the meaning of probability and give an illustration.

5. Refer to Problem 3. What is the probability that the man will not slip and fall?

6. It has been found that 10% of the marriages in a certain region result in divorce before the fifth wedding anniversary. What is the probability that a marriage in that region will (a) result in a divorce before the fifth wedding anniversary, (b) not result in a divorce before the fifth wedding anniversary?

7. It has been found during this marketing season that 15 out of every hundred tomatoes sold by a certain farmer had some bacterial canker. What is the probability that an arbitrarily selected tomato from those being offered for sale by this farmer will (a) have bacterial canker, (b) not have bacterial canker?

8. Ten percent of the students who study introductory accounting in Professor Jones's class obtain an A, 20% obtain a B, and 40% obtain a C. What is the probability that a student in Professor Jones's class will—
   (a) obtain an A?
   (b) obtain a C?
   (c) not obtain a B?
   (d) obtain less than a C?
   (e) not obtain an A or B?

9. The chances are 12 in 100 that a small-loan borrower will default on his payments. What is the probability that he will not default?

10. Illustrate how the probability that an event will occur is determined on the basis of repeated experiences.

11. A magazine salesman sells subscriptions to two monthly periodicals, *Best Stories* and *Reading Moods*, by canvassing door to door. He finds that nobody is at home in two homes out of ten; that he makes a sale in 75% of the homes where he finds somebody at home (on the average); and that when he makes a sale, about half the time it is for one of his periodicals and the other half of the time it is for the other.
   (a) What is the probability he finds somebody at home at his next call?
   (b) If he finds somebody at home on his third call this afternoon, what is the probability he will make a sale?
   (c) If he makes a sale, are the possible outcomes "subscription for *Best Stories*" and "subscription for *Reading Moods*" equiprobable? Why?
   (d) When he calls at a home, are the possible outcomes "somebody is at home" and "sell subscription for *Reading Moods*" mutually exclusive? Why?
   (e) When he calls at a home, are the possible outcomes "somebody at home" and "nobody at home" equiprobable? Why?
   (f) When he finds somebody at home, are the possible outcomes "sell," "do not sell" equiprobable? Why?

12. The personnel manager in a firm interviewed 20 applicants for an accounting position. He found that 15 (including a man called Smith) had only a moderate number of college courses in accounting but considerable job experience. The other 5 (including a man called Jones) had little job experience but considerable education in accounting. On balance, he felt that they all had equal opportunity of getting the job. What is the probability that (a) Smith gets the job, (b) Jones gets the job? How did you determine the foregoing probabilities?

13. If a clerk is as likely to be out sick on one workday as on another (Monday through Friday), what is the probability that the next time he is out sick will be a Thursday?

14. Distinguish between objective and subjective probabilities.

15. If you want to determine the probability that a particular horse will come in first in a given race on a given day, would subjective or objective probability be more suitable? Why?

16. It is desired to determine the probability that a new business will be profitable in the second year of operation or sooner.
    (a) How would you determine the required probability on an objective basis?
    (b) Would a subjective probability be preferable? Why?

17. Suppose you want to determine the probability that a lawyer will win a certain criminal case.
    (a) Would subjective or objective probability be more suitable? Why?
    (b) How would you determine the objective probability? What weaknesses would you say there are in such a determination?

18. If the probability that it will rain tomorrow is 0.25, then the probability that it will not rain tomorrow is $1 - 0.25 = 0.75$. We may compute $0.25 \div 0.75 = 1/3$, which indicates that the odds are 1 to 3 that it will rain tomorrow or the odds are 3 to 1 that it will not rain tomorrow.
    (a) In Problem 6, what are the odds that a marriage will result in a divorce before the fifth wedding anniversary?
    (b) In Problem 8, what are the odds a student will obtain C or better?

Note 10A

Additional Probability Concepts and Probability Rules

This note presents probability rules that can be used to compute probabilities based on other probabilities which are known or can be determined. We will use the symbol $P(A)$ to denote "the probability that event A will occur."

Rule A. General addition rule: Let A and B represent two events. Then,

$$P(A \text{ or } B) = P(A) + P(B) - P(A \text{ and } B)$$

Suppose an unidentified seed is planted and the probability that it will yield a yellow flower is 0.15, the probability that it will yield a long-stemmed flower is 0.10, and the probability that the flower will be both yellow and long-stemmed is 0.05. Then we may write $P(\text{yellow}) = 0.15$, $P(\text{long-stemmed}) = 0.10$, and $P(\text{yellow and long-stemmed}) = 0.05$. The probability that the flower will be yellow *or* long-stemmed is computed according to Rule A as $P(\text{yellow or long-stemmed}) = 0.15 + 0.10 - 0.05 = 0.20$.

$P(\text{yellow})$ is the probability that the flower will be yellow, no matter what its stem length. $P(\text{long-stemmed})$ is the probability that the flower will be long-stemmed, no matter what its color. Hence, both of these probabilities include the probability that the flower will be both yellow and long-stemmed. Consequently, according to Rule A, $P(\text{yellow and long-stemmed})$ is subtracted from the sum of the other two probabilities to eliminate the double inclusion of $P(\text{yellow and long-stemmed})$.

$P(\text{yellow or long-stemmed})$, computed according to Rule A, is the probability that the flower will be yellow or long-stemmed or *both* yellow and long-stemmed.

If events A and B are mutually exclusive, $P(A \text{ and } B) = 0$ in accordance with the definition of mutually exclusive events. Then, for such events Rule A becomes Rule B.

Rule B. Special addition rule: Let A and B represent mutually exclusive events. Then,

$$P(A \text{ or } B) = P(A) + P(B)$$

Suppose Jerome and Susan both apply for a position as an economist and P(Jerome gets the job) = 0.25 and P(Susan gets the job) = 0.20. Since there is only one job to be filled, the events "Jerome gets the job" (event A) and "Susan gets the job" (event B) are mutually exclusive, so $P(A \text{ and } B) = 0$. Applying Rule B, we may compute P(Jerome or Susan gets the job) = 0.25 + 0.20 = 0.45.

Rule B can be extended to any number of mutually exclusive events. For example, for three events we have the following:

Rule B Extended: Let A, B, C represent mutually exclusive events. Then,

$$P(A, B, \text{or } C) = P(A) + P(B) + P(C)$$

In the foregoing illustration, if Ethel also applied for the position and P(Ethel gets the job) = 0.30, we may compute P(Jerome, Susan, or Ethel gets the job) = 0.25 + 0.20 + 0.30 = 0.75.

Rule C. General multiplication rule: Let A and B represent two events. Then,

$$P(A \text{ and } B) = P(A) \times P(B \mid A)$$

The symbol $P(B \mid A)$ denotes "the probability that B will occur given that A has occurred." Statisticians call $P(B)$ the *unconditional probability* that B will occur and $P(B \mid A)$ the *conditional probability* that B will occur; the condition being that A has already occurred.

Suppose two cards are selected by a blindfolded person from a well-shuffled deck of 52 playing cards. What is the probability that the first card will be a king and the second will be a queen? Assume that the first card drawn is *not replaced* before the second card is selected. According to Rule C we may write the following:

$$P(\text{king first and queen second}) = P(\text{king}) \times P(\text{queen} \mid \text{king})$$

When the first card is to be selected, there are 52 mutually exclusive, equiprobable possible outcomes (because the deck contains 52 cards). Four of these 52 cards are kings, so the chances are 4 in 52 that a king will be selected, or P(king) = 4/52. This follows from our discussion, in chapter 10, of computation of probabilities for mutually exclusive, equiprobable events.

When the second card is to be selected, the first card is no longer in the deck and we assume that it is a king (as desired), so only 51 cards remain in the deck. Then the second selection has 51 mutually exclusive, equiprobable possible outcomes and 4 of these are queens. Consequently, the chances are 4 in 51 that the second card is queen, or P(queen|king) = 4/51. Then, applying Rule C we compute the following:

$$P(\text{king first and queen second}) = \frac{4}{52} \times \frac{4}{51} = \frac{16}{2652} = 0.006$$

If we want to compute the probability that a queen is selected first and then a king, following the same reasoning as before, P(queen) = 4/52 and P(king|queen) = 4/51. Then, applying Rule C P(queen first and king second) = 4/52 × 4/51 = 0.006.

What is the probability that a king and a queen are selected, regardless of which is selected first? We must take into consideration that this could occur in two possible ways, king first and queen second (event A) or queen first and king second (event B). Events A and B are mutually exclusive, since only one of these could occur. Then applying Rule B, where $P(A \text{ or } B)$ is more conveniently written as P(king and queen), we have the following:

$$\begin{aligned} P(\text{king and queen}) &= P(\text{king first and queen second}) + P(\text{queen first and king second}) \\ &= 0.006 + 0.006 = 0.012 \end{aligned}$$

Rule C may be extended to any number of events. For example, for three events we have the following:

Rule C Extended: Let A, B, C represent three events. Then,

$$P(A, B, \text{and } C) = P(A) \times P(B \mid A) \times P(C \mid A \text{ and } B)$$

The symbol $P(C \mid A \text{ and } B)$ denotes the conditional probability that C will occur given that both A and B have occurred.

Suppose three cards are to be selected by a blindfolded person from a well-shuffled deck of 52 playing cards. What is the probability that the first card is a king, the second is a queen, and the third is a jack? Assume that cards selected are *not replaced* to the deck before the next card is selected. Following the same reasoning as before, $P(\text{king}) = 4/52$, $P(\text{queen}|\text{king}) = 4/51$, and $P(\text{jack}|\text{king and queen}) = 4/50$. Then, applying Rule C Extended, we have the following:

$$P(\text{king first, queen second, jack third}) = \frac{4}{52} \times \frac{4}{51} \times \frac{4}{50}$$

$$= \frac{64}{132{,}600} = 0.0005$$

Therefore, the chances are 5 in 10,000 that the three cards selected will be king, queen, jack, *in that order*. If we wanted to determine the probability that these three cards will be selected in some other order, say queen, king, jack, the probability would be the same.

If it is desired to compute the probability that king, queen, jack are selected regardless of order, then we must take into consideration all possible orders in which these cards could be selected, as follows:

1. king, queen, jack
2. king, jack, queen
3. queen, king, jack
4. queen, jack, king
5. jack, king, queen
6. jack, queen, king

If three cards are selected only once, then only one of these orders (called *permutations*) could occur. Therefore, these six permutations represent six mutually exclusive possible outcomes and, as previously noted, the probability that a specified one will occur is 0.0005. Then, applying Rule B Extended, $P(\text{king, queen, jack}) =$ sum of the probabilities of the six mutually exclusive events $= 6 \times 0.0005 = 0.003$.

In some problems there is no need to take permutation into consideration. For example, suppose two cards are selected from a deck of 52 playing cards, without replacement. What is the probability that two kings are selected? Since both cards are kings, permutation is not involved and the desired probability is $P(\text{king and king}) = P(\text{king}) \times P(\text{king}|\text{king}) = 4/52 \times 3/51 = 12/2652 = 0.005$.

$P(\text{king}) = 4/52$ for the first card selected because the deck contains 52 cards, including 4 kings. However, when the second card is selected, the deck contains only 51 cards and it is assumed that the first card selected was a king, so only 3 kings remain in the deck. Therefore, for the second card selected, $P(\text{king}|\text{king}) = 3/51$.

Suppose two cards are selected from a deck of 52 playing cards. Now assume that the first card selected is *replaced* to the deck before the second card is selected. The events "king first" and "queen second" are *independent events* because the occurrence or nonoccurrence of the first event has no effect on the probability that the second event will occur. $P(\text{king first}) = 4/52$, as before, and $P(\text{queen second}) = 4/52$ because the first card was replaced and the deck contains 52 cards when the second card is selected.

Similarly, if a coin is tossed twice, the events "heads first" and "tails second" are independent because the outcome of the first toss has no effect on the probability that the second toss will result in a tail. Events which are not independent are called *dependent events.*

Rule D. Special multiplication rule: Let A and B represent independent events. Then,

$$P(A \text{ and } B) = P(A) \times P(B)$$

In the foregoing illustration where two cards are selected *with replacement*, P(king first and queen second) $= (4/52) \times (4/52) = 16/2704 = 0.0059$. Similarly, P(queen first and king second) $= 0.0059$. Finally, the probability that a king and a queen are selected, with no interest in permutation, is computed by applying Rule B to obtain P(king and queen) $= 0.0059 + 0.0059 = 0.012$.

Rule D can be extended to more than two independent events. For three independent events, we may write the following:

Rule D Extended: Let A, B, C represent three independent events. Then,

$$P(A, B, \text{and } C) = P(A) \times P(B) \times P(C)$$

As before, Rule D Extended is applied to compute the probability of a specified permutation of the events.

11

Random Sampling

11.1 Representative Samples

Suppose a shirt manufacturer wants to know how many of the 15 men in his shipping department like a new style shirt being considered for production. All he has to do is ask each man directly or call them all together to get their reactions. In this case, the population (the opinions of the 15 men) is small and the information to be obtained is simple. On the other hand, if the manufacturer wishes to obtain such information from all the men in the country, the population would be very large and the information required would be difficult and costly to obtain.

Suppose a social service agency maintains a file card for each family who received aid through their agency. If it is desired to obtain the proportion of the families helped who are Spanish-American families, it is practical to determine this by examining the information in each card only if the number of cards in the file is reasonably small. If the number of cards is large (say 1,000 or more), it may not be practical to examine all the cards to obtain the required information. Moreover, if the information to be determined from each card is detailed (e.g., size of family, age of family head, dates of aid provided, type and amount of aid provided, etc.), it would be practical to collect this information from the entire population of cards only if the number of cards were quite small.

Typically, it is impractical and costly to collect data from an entire population and it takes too long to produce the required information. The reasonable approach is to collect *sample data*. Statistical analyses based on sample data, if properly carried out, may be expected to provide useful and adequate information about the population.

As a final illustration, if a producer of flashlight batteries wishes to determine the average life of a new type of battery, he has no choice but to use a sample of the batteries for this purpose. If he used all the batteries to determine average life, he would have to use all of them until they are no longer useful. Then, he would have none left to sell.

When we speak about sample data, we do not mean any type of sample. It is certainly not true that any type of sample provides a useful basis for estimating population characteristics. For example, if the shirt manufacturer decides to use the opinions of the 15 men in his shipping department as a sample of the opinions of all men in the country, would such a sample provide a useful and adequate estimate for the population of all men in the country? If a clerk in the social service agency were asked to grab a handful of file cards and determine the required information from this small set of cards, could these findings be used to estimate the characteristics of all families whose cards are in the file? More specifically, are these samples *representative* of the population from which they were selected? This is the important consideration in sample selection.

A *representative sample* is one which reflects the characteristics of the population sampled in true population proportions.

The only way to insure that a sample is representative is to know the population thoroughly and to select the sample carefully based on this knowledge. If complete information about a population is not available, *a representative sample can never be guaranteed.* Could the samples in the foregoing illustrations be considered representative samples?

When complete information about a population is not available and where little is known about the population, a representative sample is more likely to be obtained if a *random selection procedure* is used than if the sample is selected in any other way. Such samples are called *random samples.* Statisticians always use random samples to prepare estimates of population characteristics. When we speak of samples in this text, we will always mean random samples.

11.2 How to Select a Random Sample

A random sample cannot be obtained by haphazard selection from a population. A random sample can be obtained only by applying a random selection procedure.

Random selection is easier to understand if we think in terms of sampling marbles from a large pot. Suppose we wish to select a random sample of 25 retail stores from the 500 retail stores in a city. Let us imagine that we have 500 marbles in a large pot and each marble has printed on it the address of one of the stores in the population. Then, our problem is to select a random sample of 25 marbles from the pot.

This is accomplished as follows: Stir the marbles in the pot vigorously and thoroughly and, while blindfolded, select a marble. Then, without replacing the marble to the pot, stir the marbles again vigorously and thoroughly and, while blindfolded, select a second marble. Repeat this procedure until 25 marbles are selected. The stores whose addresses are on the selected marbles make up a random sample of 25 stores from the population of 500 retail stores. This method of selection is a random selection procedure because each time a marble is selected, each marble in the pot has an equal chance of being the one chosen. The vigorous and thorough stirring of the marbles in the pot insures *equal probability (chances) of selection* for all the marbles in the pot.

There are many combinations (samples) of 25 retail stores which could be selected from the population of 500 stores. If the foregoing random selection procedure is followed, then each possible sample (combination) of 25 stores has an equal chance of being the sample actually selected.

In the foregoing sampling procedure, it is specifically stated that the selected marbles are not to be returned to the pot. This is the usual method for selecting a random sample. Statisticians call this *sampling without replacement.* Occasionally, samples are selected with replacement. (This means that the marble selected would be returned to the pot and could be selected again.) When *sampling with replacement*, a sample of 25 stores may be made up of some stores which are considered more than once. For example, the sample may include store A twice, store B three times, and 20 other stores. Such a sample is also a statistically acceptable random sample. However, nearly always, sampling is without replacement, as previously discussed.

Of course, the use of marbles to select a random sample is not practical and merely serves to present an example of the random selection method. A more practical method is use of a *table of random numbers* or, as it is also called, a *table of random digits.* Table 11.1 is such a table. The random digits presented were taken from the last page of a publication which contains 105,000 random digits.

The digits that go into a table of random numbers may be determined by a randomizing procedure not unlike our procedure for selecting marbles from a pot. Ten marbles with the ten digits 0, 1, 2, . . . 9 appearing one on each marble would be needed for this. However, for this purpose, sampling would be with replacement. There are more complicated methods that are also used for constructing a table of random numbers. However, we need not concern ourselves with these.

The first step in using a table of random digits to select a sample is to identify each item in the population with a number. Say we wish to sample a population of 750 nursing homes in a region. Each nursing home must be assigned a number from 001 to 750. Notice that the first number assigned is 001, not 1. Since the highest number to be assigned (750) has three digits, each number must be expressed as a three-digit number.

These numbers may be assigned in any convenient way. One way would be to list the nursing homes and assign the numbers in order, column by column. Or, the nursing homes might be spotted on a map with the identifying number for each home posted on the map. No matter which method is used, it is *absolutely necessary* that it be firmly decided, *in advance*, which number identifies which nursing home in the population.

Next it must be decided exactly where to begin in the table of random numbers and how three-digit numbers are to be formed. Notice that the random numbers in Table 11.1 are grouped into blocks of five columns and five rows. This is to aid the user of the table. The column headings (1) through (7) identify the seven blocks in each row. The line numbers identify the rows. (In the random number publication from which Table 11.1 was constructed, the first line number on the first page is 1 and there are actually 14 blocks per row.) We may decide arbitrarily to begin with

Table 11.1

Table of Random Numbers

Line \ Col	(1)	(2)	(3)	(4)	(5)	(6)	(7)
1476	05835	36457	15516	59987	43545	84159	39716
1477	89456	73108	57063	46675	89930	85054	75680
1478	44830	07509	23981	71725	10318	27204	23296
1479	39864	47322	31251	82524	45643	31038	20455
1480	94135	03380	81698	48694	52326	75025	50761
1481	69240	52865	81175	30481	17890	64355	96345
1482	79725	53403	89131	02804	26309	78157	29432
1483	71198	93772	90685	29615	42387	72649	38219
1484	09108	41317	42548	80675	32823	12150	66274
1485	16637	65387	79852	38250	44973	45612	44938
1486	63847	13687	45157	64816	50626	31236	34127
1487	33924	92613	47288	49738	37532	82272	44805
1488	72094	93704	50741	90556	72074	27652	85568
1489	10274	87910	51552	22331	60297	22908	96064
1490	45853	62407	77906	62399	95568	32481	10186
1491	90163	21938	53874	51545	72222	78362	48573
1492	61992	66865	35404	71123	23733	88910	45229
1493	13829	31033	62790	91935	56721	21673	52965
1494	09775	54186	21927	51460	26755	22921	50617
1495	09592	17859	56164	45701	20939	92689	69421
1496	42639	84884	69790	90784	34940	18891	69850
1497	21655	81382	78529	20469	22066	96281	97864
1498	80111	69552	54281	44357	98385	21425	72915
1499	09761	22879	57602	28894	07030	91155	95164
1500	47808	83766	69073	93127	76351	14177	05713

SOURCE: "Table of 105,000 Random Digits," U.S. Interstate Commerce Commission, 1955, a portion of page 30.

the first column on line (or row) 1481, so that our first random digit is 6 (find this in the table). Then, we may agree to form three-digit numbers by moving across this row toward the right, forming the random numbers 692, 405, 286, etc.

Or, we may decide to choose the three-digit number in the first three columns of line 1481, 692, and continue to form three-digit numbers by going down these three columns. Then, we obtain the additional random numbers 797, 711, 091, etc. It does not matter which method for entering the table and forming numbers is chosen. However, it is absolutely essential that the procedure is determined and carefully specified before looking into the body of the random number table. This will insure that the user has no opportunity to influence the numbers selected in any way.

Finally, random numbers are copied out of the random number table until the desired number of *acceptable numbers* are obtained. If a sample of 50 nursing homes is desired, then 50 acceptable numbers should be copied out of the table. *Acceptable numbers* are those that are consistent with the population size and the sampling plan. In our illustration, the population of nursing homes was assigned numbers 001 through 750. Therefore, any random number read out of Table 11.1 that is larger than 750 is not acceptable and should be discarded. If sampling is to be without replacement (the usual situation), then a number is acceptable only the first time it is selected. Then, the 50 nursing homes which are identified by the 50 acceptable numbers make up the random sample required.

11.3 Types of Random Samples

There are many types of random samples which could be selected from a population. A plan for selecting a sample and computing the required estimates of population characteristics is called a *sampling design* or a *sampling plan*. It is useful to briefly describe some of the sampling plans frequently used by statisticians.

The random sample we have described is called a *simple random sample*. Sometimes a *systematic random sample* is selected. In our nursing home population, a systematic random sample would be selected by first identifying each home with a number from 1 through 750, as before. Then, some interval, say 10, would be specified and a number in this interval (from 01 to 10) would be selected on a random selection basis. This might be accomplished through a table of random numbers. After it is decided where to enter the table and how to form two-digit numbers, the first acceptable two-digit number would be selected.

Say the selected number is 06. Then, the nursing home numbered 6 in the population would be the first home selected for the sample. Then, homes numbered $06 + 10 = 16$, $16 + 10 = 26$, $26 + 10 = 36$, etc., would be selected until 750 were reached, but not exceeded. The nursing homes identified by these numbers (6, 16, 26, etc.) would then make up a systematic random sample of nursing homes.

A *stratified random sample* design is also often used. Suppose you wish to estimate the average monthly rental for the 1,500 apartments put on the market in a city last year and a stratified random sample is to be used for this purpose. To set up a stratified random sample, you first divide the population of apartments into relatively homogeneous groups (from the point of view of rents charged per apartment). For example, we could set up the three groups: apartments in high-income areas, apartments in middle-income areas, and apartments in low-income areas. These apartment groups represent three subpopulations. The monthly rentals will be more alike in a subpopulation than for the entire population of 1,500 apartments. This will permit the statistician to use a smaller sample and the sampling project could be carried out at less cost, in a shorter time.

In stratified random sampling, the subpopulations are called *strata* (each subpopulation is a *stratum*). Then, a simple random sample is selected from each stratum separately (using a table of random numbers and following the procedure previously discussed). Finally, the average monthly rental for the entire population of 1,500 apartments is estimated on the basis of the data collected in the random samples from all the strata.

Cluster sampling is another type of sample frequently used. In our population of 1,500 apartments, we may designate the apartments in a single building as a *cluster* (a group of apartments). Then, a cluster sampling plan requires selection of a random sample of clusters (buildings). The estimate of average monthly rental is based on the apartments included in the sample of clusters. A simple random sample of clusters is obtained by assigning a number to each cluster and then employing a table of random numbers as previously explained.

11.4 Some Pitfalls in Sampling

An important aspect of design of sampling plans is precise definition of the population to be sampled and insurance that the sample is selected from the specified population. This may appear to be a simple matter to those not trained in statistical sampling. However, this phase of designing a sample is full of pitfalls.

Suppose it is desired to select a sample of voters in Oakland, California to estimate the proportion who favor candidate Smith. A simple and *apparently* satisfactory method is to select a sample of names from the city telephone directory. Actually, such a sample is not adequate because low-income voters and others who might not have a telephone will have no chance of being selected for the sample. Similarly, new telephone subscribers may not yet be listed and will have no chance of being selected. A random sample selected from the directory would be a random sample of telephone subscribers who are listed in the directory and not a random sample of voters in Oakland.

Say it is desired to sample accounting firms in Cleveland to estimate the average income of accountants. It must be specifically stated what is meant by "accounting firm." Are one-man firms to be included? Furthermore, what is meant by "accountant"? Are part-time accountants to be included? Are employed accountant-trainees to be included?

Sometimes the population to be sampled is purposely restricted to reduce survey costs and the time required to perform the survey. Suppose it is desired to sample public high schools in the nation to determine average size of class. It is decided to restrict the population of high schools only to those in cities of 500,000 or more population. Such a restriction will reduce survey costs for data collection and data compilation. However, it should be realized that the estimates prepared from this *restricted survey* are to be interpreted as applying only to the *restricted population* of high schools from which the sample was selected and not to the entire population of high schools. The sample survey findings can be interpreted as applying to the entire population of high schools on a judgmental basis only, not on a statistical basis.

Review

Concepts and Terminology

Acceptable numbers
Cluster
Cluster sample
Sampling plan
Sampling with replacement
Sampling without replacement

Concepts and Terminology *(cont.)*

Equal probability of selection
Randomizing procedure
Random sample
Random selection procedure
Representative sample
Restricted population
Restricted survey
Sampling design
Simple random sample
Statistical sampling
Strata
Stratified random sample
Stratum
Systematic random sample
Table of random digits
Table of random numbers

Problems

1. What is meant by a representative sample?
2. A union official polled the members of his local union and found that 90% favored Smith for president of the national organization. He stated, based on this finding, that Smith was the clear favorite of all union members in the country. Do you agree with this statement? Why?
3. A man stands at a busy corner in a large city during the lunch hour on a business day and asks the first 100 men who pass him whether they use an electric shaver. Suppose he finds that 70% of these men use electric shavers. Should he conclude that 70% of all men in the country use electric shavers? Why?
4. A sample of TV tubes produced by a certain firm is tested to determine the length of life of TV tubes produced in the country today. Would you consider this sample representative of the population? Why?
5. A sample of college seniors is polled to determine the attitude of the voters in the nation toward a tax bill being discussed in Congress. Would you consider this to be a representative sample? Why?
6. Distinguish between sampling with replacement and sampling without replacement. Which is most often used?
7. A class in business management is made up of 25 students. Explain how you would go about selecting a random sample (without replacement) of five students, using a table of random numbers (such as Table 11.1).
8. The following account balances (in dollars) were listed for the current accounts of a firm:

2050	1905	1480	2111	1610	915
1836	2005	1942	1814	1825	809
1481	1000	2015	703	1105	1202
3600	800	1555	920	1962	1861
1002	732	1603	3015	2163	993

Using Table 11.1, select a random sample (without replacement) of the size indicated and state each step in your selection procedure.

(a) A sample of 5 balances. Start with the first digit on line 1481 and form random numbers by reading digits across this line and continuing line by line.

(b) A sample of 8 balances. Start with the digit in the fifth column, line 1478, and form random numbers by reading digits down this column and continuing column by column.
(c) A sample of 10 balances.

9. In a table of random digits used to select a random sample, what do we mean by "acceptable numbers"? Give an example.

10. What is meant by a "sample design"? What is another name for this?

11. Describe briefly what is meant by:
(a) simple random sample
(b) systematic random sample
(c) stratified random sample
(d) cluster sample

12. Suppose it is desired to select a sample of physicians to estimate the average yearly income of a doctor. Would it be satisfactory to select the sample from the doctors listed in the medical directory? Why?

13. A sample of exporters is selected from the membership list of a national exporters' association in order to estimate the average annual volume of exports of an exporter. What is the population of exporters to which the sample results could be generalized? How can the sample findings be generalized to all exporters in the country?

14. A collection of 60 petri dishes containing nutrient agar showed the following amounts of agar in each dish (in milliliters):

75	20	15	35	70	40	18	50	90	65
15	70	20	35	30	45	50	85	30	70
65	80	25	30	20	75	60	45	40	50
15	50	65	20	55	25	20	50	30	30
65	60	15	30	45	35	25	40	35	16
55	80	60	40	15	95	45	20	25	60

Using Table 11.1, select a random sample of petri dishes of the size indicated, without replacement, and state each step in your selection procedure.
(a) A sample of size 10. Enter Table 11.1 on line 1489, first column in block (3).
(b) A sample of size 15. Enter Table 11.1 on line 1479, last column in block (5).
(c) A sample of size 20.

15. Refer to Problem 14. Using 4 as the sampling interval, select a systematic random sample of petri dishes. State each step of your sample selection procedure.

16. Refer to Problem 8. Consider each column of balances as a cluster. Select a random sample of three clusters. Indicate each step of your sample selection procedure.

17. Refer to Problem 14. Consider each row of petri dishes as a stratum. Select a stratified random sample of petri dishes made up of three dishes selected from each stratum.

12

The Risks of Sampling

12.1 Useful and Adequate Estimates

Let us consider a population of 5 weights (in pounds): 5, 10, 15, 20, 25. The arithmetic mean of this population is 15 pounds and the standard deviation is about 7 pounds. Let us suppose that we do not know the population mean and wish to estimate it based on a random sample of 2 weights (a sample of size 2). Of course, random selection of 2 weights out of 5 could result in any 2 weights in the population being selected. What are the possibilities?

Table 12.1 shows all the samples of size 2 which could possibly be selected from this population. This table also shows the mean computed from each sample. According to this table, if the random sample happened to be the weights 5 pounds and 10 pounds, then the sample mean would be 7.5 pounds and the statistician would estimate the population mean weight as 7.5 pounds. On the other hand, if the weights 5 pounds and 15 pounds made up the random sample, the sample mean would be 10 pounds and he would estimate the population mean weight as 10 pounds.

Table 12.2 organizes the possible sample means obtainable from a sample of two weights and shows the number of ways each mean could be obtained (frequency). For example, this table shows that a sample mean of 7.50 pounds has a frequency of 1 (since it could be obtained in only one way if the sample is made up of the weights 5 pounds and 10 pounds). On the other hand, a sample mean of 12.50 pounds could be obtained in two ways (if the sample is made up of the weights 5 and 20 pounds or 10 and 15 pounds).

It is interesting to notice from Table 12.2 that the sample mean, which is used as an estimate of the population mean (15 pounds), could be as low as 7.50 pounds and

Table 12.1

All Possible Samples of Two Weights Selected from a Population of Five Weights

Weights in Sample (Pounds)	Sample Mean (Pounds)
5,10	7.50
5,15	10.00
5,20	12.50
5,25	15.00
10,15	12.50
10,20	15.00
10,25	17.50
15,20	17.50
15,25	20.00
20,25	22.50

Table 12.2

Distribution of Sample Means Obtainable from Samples of Two Weights

Sample Mean (Pounds)	Frequency	Proportion
7.50	1	$\frac{1}{10}$
10.00	1	$\frac{1}{10}$
12.50	2	$\frac{2}{10}$
15.00	2	$\frac{2}{10}$
17.50	2	$\frac{2}{10}$
20.00	1	$\frac{1}{10}$
22.50	1	$\frac{1}{10}$
Total	10	

as high as 22.50 pounds. Does this mean that random samples are not useful as a basis for estimating population characteristics? Not at all! Notice that 2 out of the 10 possible samples yield a sample mean of exactly 15 pounds, so the estimate of the population mean would be exactly right. The last column in Table 12.2 shows that 2/10 or 0.20 of the time a random sample of two weights will produce such an estimate of the population mean. The proportions shown in this table are to be interpreted as *probabilities*. Then, the probability is 0.20 or 20 chances in 100 that a random sample of two weights will result in a perfect estimate of the population mean weight.

However, nobody really expects that an estimate based on a sample will be exactly correct. When estimates are computed from sample data, the expectation is that they will be *useful* and *adequate*. In other words, the expectation is that the *sampling error will be small*. Notice from Table 12.2 that if we are satisfied with an estimate of the population mean that is off either way by no more than 2.5 pounds, sample means of 12.50 pounds, 15.00 pounds, and 17.50 pounds are adequate. Note that 6 samples out of 10 provide such estimates of the population mean, so that the probability is 0.60 or 60 chances in 100 that the random sample will provide a useful and adequate estimate.

If we are satisfied with an estimate that is off by no more than 5 pounds either way, notice from Table 12.2 that 8 samples out of 10 provide estimates between 10 pounds and 20 pounds; so the probability is 0.80 or 80 chances in 100 that a useful and adequate estimate will be obtained. Then, there is a risk of only 20 chances in 100 that an inadequate estimate will be produced by a sample of size 2.

When using a sample to produce an estimate of a population characteristic, we must always be aware that there is a risk of obtaining a poor estimate. In practice, the risk of a poor estimate is always less than indicated in our illustration. The risk is usually only 5 chances in 100 and, if desired, it could be reduced to even a lower risk (perhaps, only 2 chances in 100 or even 1 chance in 100).

12.2 Controlling the Risks

How can a statistician control the risk that a sample will provide an inadequate estimate? There are two factors which determine this risk: the *amount of variation in the population* (measured by the standard deviation), and the *size of the sample* used to compute the estimate. There is usually nothing a statistician can do to reduce the variation in the population. However, often he can reduce its effect (for example, by use of a stratified sampling plan). The greater the population variation, the greater the risk of an inadequate estimate. The statistician can and does control the sample size. The larger the sample size, the smaller the risk that an inadequate estimate will be obtained.

Suppose we increase the sample size in our illustration from two weights to three weights. Table 12.3 shows all the possible samples which could be obtained and Table 12.4 shows the distribution of sample means for these samples.

Compare Table 12.4 with Table 12.2. Notice that for samples of three weights (Table 12.4), sample means range from a low of 10.00 pounds to a high of 20.00 pounds; so it is a certainty (100 chances in 100) that the estimate of the population mean weight will not be off by more than 5 pounds either way if a random sample of size 3 is used. As previously noted (and shown in Table 12.2), if a sample of size 2 is used, there are 80 chances in 100 that the sampling error will be no more than 5 pounds either way.

Table 12.3

All Possible Samples of Three Weights Selected from a Population of Five Weights

Weights in Sample (Pounds)	Sample Mean (Pounds)
5,10,15	10.00
5,10,20	11.67
5,10,25	13.33
5,15,20	13.33
5,15,25	15.00
5,20,25	16.67
10,15,20	15.00
10,15,25	16.67
10,20,25	18.33
15,20,25	20.00

Table 12.4

Distribution of Sample Means Obtainable from Samples of Three Weights

Sample Mean (Pounds)	Frequency	Proportion
10.00	1	$\frac{1}{10}$
11.67	1	$\frac{1}{10}$
13.33	2	$\frac{2}{10}$
15.00	2	$\frac{2}{10}$
16.67	2	$\frac{2}{10}$
18.33	1	$\frac{1}{10}$
20.00	1	$\frac{1}{10}$
Total	10	

12.3 Sampling Distributions

The distributions of the sample mean shown in Table 12.2 and Table 12.4 are called *sampling distributions*. More specifically, they are called *sampling distributions of the mean*. Sampling distributions are the most important distributions in statistics. Such distributions are the basis for statistical thinking whenever statisticians produce population estimates based on random samples. Some sampling distributions are very complicated. However, the basic ideas of sampling distributions are similar to those brought out in our discussion.

Review

Concepts and Terminology

All possible samples from a population
Proportions as probabilities
Risks of sampling
Sample size
Sampling distribution
Sampling distribution of the mean
Sampling error
Useful and adequate population estimates

Problems

1. "You can never expect to obtain a perfectly accurate measure of a population mean based on sample data." True or false? Explain your answer.
2. "I don't want to use sample data to estimate the population mean because it could be way off." Do you agree? Comment.
3. "The risk that a random sample will not provide an adequate estimate of a population mean can be controlled by the statistician." Explain.
4. A population of five flats (shallow wooden boxes) containing a mixture of sterilized and nonsterilized muck soil with damping-off fungi has 100 pea plants per flat. A count was made of the number of plants in a flat with height above 1.5 centimeters and found to be 6, 2, 9, 10, 3.
   (a) Compute the population mean number of plants per flat with height above 1.5 centimeters.
   (b) Construct a table (A) similar to Table 12.1 for possible samples of size 2 selected from the population of 5 counts.
   (c) Construct a table (B) similar to Table 12.2 using the data in Table (A).
   (d) What is the probability that a random sample of 2 counts will provide an estimate of the population mean count that is off by more than one either way?

(e) What is the probability that a random sample of 2 counts will provide an estimate of the mean that is off by 2 or more either way?

5. Refer to Problem 4.

(a) Construct the sampling distribution of the mean for samples of size 3.

(b) What is the probability that a random sample of 3 counts will provide an estimate of the population mean which is off by more than one, either way?

(c) What is the probability that a random sample of 3 counts will provide an estimate of the population mean which is off by 2 or more either way?

6. The secretary in a firm of corporate lawyers recorded the number of trial days spent on five cases as 6, 7, 5, 3, 2.

(a) Compute the mean number of trial days for this population of cases.

(b) Construct the sampling distribution of the mean for samples of size 2 selected from this population.

(c) Construct the sampling distribution of the mean for samples of size 3 selected from this population.

(d) If the sample mean is used as an estimate of the population mean, what is the probability that your estimate will be between 4 and 6 trial days if a random sample of 2 cases is used? A random sample of 3 cases?

(e) What is the probability that your estimate of the population mean will be between 4.5 and 5.5 trial days if a random sample of 2 cases is used? A random sample of 3 cases?

13

How the Statistician Sees It

The sampling distribution concept introduced in the previous chapter provides the base on which the statistician formulates his statistical thinking. More specifically, the sampling distribution involved when a population characteristic (such as the mean) is estimated based on a random sample provides the basis for estimating *how good* a particular estimate is and for measuring the *risk* that the estimate is inadequate.

The notion of sampling risks and the concept of a sampling distribution were discussed in the previous chapter in terms of sampling from a very small population. Though unrealistic, use of a very small population provided a simplified basis for introducing these concepts.

Typically, the statistician is concerned with sampling from large populations. He develops the sampling distributions involved in his problem on the assumption that he is sampling from a *theoretical population.* Then, he determines how good his estimates are and how much risk is involved based on an appropriate *theoretical sampling distribution.* Two theoretical distributions frequently used by statisticians are the *normal distribution* and the *binomial distribution.*

13.1 Normal Distribution

As an aid to understanding the *normal distribution*, a review of the histogram and distribution shapes would be helpful. Recall that the histogram is used to display the shape of the distribution of a set of data. Also, the histogram (and the distribution of the data) may be approximated by a smooth, continuous curve.

Suppose we have a large set of weights (in pounds) and that the mean weight is 100 pounds, and the standard deviation is 20 pounds. Suppose the weight data are organized into a frequency distribution and the length of each class interval is 20 pounds (*the standard deviation*). The histogram in Figure 13.1(a) was constructed for this frequency distribution. The arrow on the baseline of this histogram is at the mean (100 pounds) and there are three classes (bars) on each side of the mean. The width of each bar is 20 pounds (*the standard deviation*).

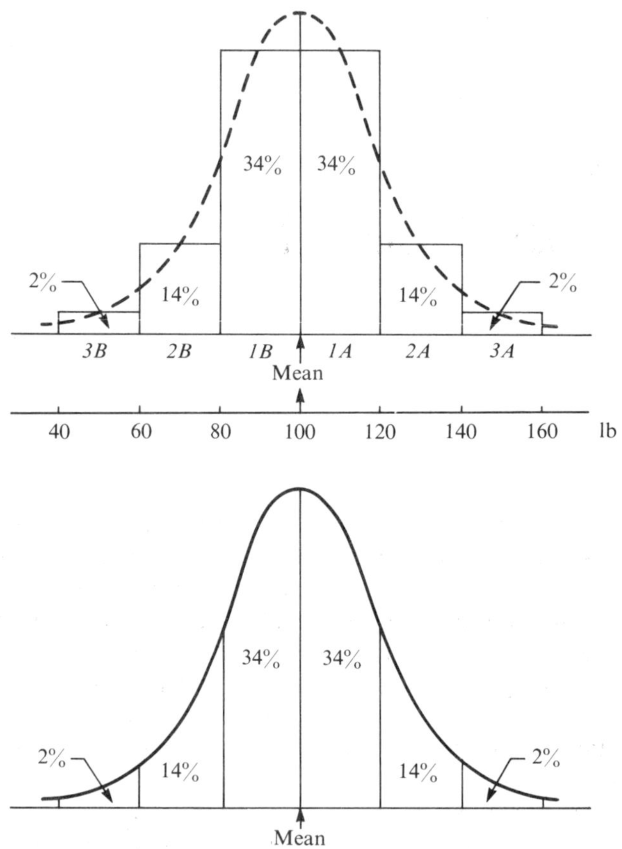

Figure 13.1

The Normal Distribution

The bars in the histogram are identified as *1A*, *2A*, *3A*, and *1B*, *2B*, *3B*. The percentage shown in a bar indicates the percentage of the weights which fall into each class. More specifically:

(a) Bar *1A* indicates that 34% of the weights are between 100 lb (the mean) and 100 + 20 = 120 lb.

(b) Bar *2A* indicates that 14% of the weights are between 120 lb and 120 + 20 = 140 lb.

(c) Bar *3A* indicates that 2% of the weights are between 140 lb and 140 + 20 = 160 lb.
(d) Bar *1B* indicates that 34% of the weights are between 100 − 20 = 80 lb and 100 lb.
(e) Bar *2B* indicates that 14% of the weights are between 80 − 20 = 60 lb and 80 lb.
(f) Bar *3B* indicates that 2% of the weights are between 60 − 20 = 40 lb and 60 lb.

The dashed-line continuous curve which approximates the shape of the histogram in the figure is a *normal curve* or a *normal distribution.* The normal distribution always has the percentages in each of the class intervals (where class length equals the standard deviation) as shown. The curve in the lower panel of Figure 13.1 shows the normal distribution without the underlying histogram. The arrow on the base line of this curve identifies the position of the mean.

Many populations encountered in practice may be approximated by the normal distribution. For example, the time required by a worker to do a job, determined for a large number of workers, may well show a distribution which is approximately normal. Similarly, the number of defective bolts produced by a machine per hour, the number of errors made by typists, the grade averages obtained by college seniors, the weight of 10-year-old boys, and the height of matured men have distributions which may be approximated by the normal distribution.

Say we know that hourly production per worker for an industry has a mean of 50 units per hour and a standard deviation of 5 units. If production per worker is approximately normal in distribution, then we may estimate that 34% of the workers have a productivity between 50 units per hour and 50 + 5 = 55 units per hour. This is indicated in bar *1A* in Figure 13.1 (a) or in the corresponding portion of the normal distribution in Figure 13.1(b).

Many sampling distributions are well approximated by the normal distribution. This is the reason that the normal distribution is the most important theoretical distribution in statistics. The statistician bases his statistical thinking on this distribution in the great majority of his problems. For example, when estimating a population mean based on a sample, the statistician can often use the normal distribution to approximate the sampling distribution related to his problem. Then, based on the characteristics of the normal distribution (such as the percentages shown in Figure 13.1), he can say how good his estimate is and how much risk there is that his estimate is inadequate. (**Note 13A** at the end of the chapter presents a more technical discussion of the normal distribution. **Note 13B** presents a more technical discussion of the sampling distribution of the mean.)

13.2 Binomial Population

A *binomial population* is a population made up of two categories. A population of adults identified only as male or female and a population of grades identified as pass or fail are binomial populations. In other words, if a population of observations (prices, weights, grades, divorce rates, wages, etc.) is classified into two categories, we have a binomial population.

Whenever a statistician is interested in a *proportion*, he thinks in terms of a binomial population. Suppose he is interested in the proportion of defective bolts produced by a machine. Then, from the point of view of his interest, the bolts produced by the machine fall into two categories, defective and not defective. Suppose a statistician samples the farms in the population in order to estimate the proportion which gross under $10,000 a year. Then, from his point of view, the population of farms falls in two categories, those which gross under $10,000 a year and all other farms.

The items in the two categories of a binomial population are usually coded 1 and 0. The items in the category of interest are coded 1, the other items are coded 0. For example, if a statistician samples a population of juvenile delinquents in order to estimate the proportion who come from low-income families, juveniles from low-income families are coded 1 and all the other juveniles in the population are coded 0. Then, the population of juveniles becomes a population of 1s and 0s. Actually, statisticians often refer to a binomial population as a *population of 1s and 0s*.

When sampling from a binomial population, we expect that the sample will include items from both categories. Say a sample of 50 housewives is selected from all the housewives in a town in order to estimate the proportion who prefer detergent A. We would expect that the sample would contain some housewives from category 1 (prefer detergent A) and some from category 0 (other housewives).

Suppose a statistician wishes to select a sample of 5 primary-grade school teachers in a city to estimate the proportion of the population of such teachers who do not have a teaching license. Category 1 is "primary-grade teacher with no license" and category 0 is "other primary-grade teacher." Clearly, the sample of 5 teachers could include anywhere from 0 to 5 teachers from category 1, as indicated in the first column of Table 13.1. Then, the *proportion* of the sample of 5 teachers who are from category 1 could vary from 0 to 1.00, as shown in the second column of the table.

Table 13.1

Possible Samples of Five Teachers When Sampling from a Binomial Population

Possible Numbers of Teachers in Category 1	Possible Proportions of Teachers in Category 1
0	$0 = 0$
1	$\frac{1}{5} = 0.20$
2	$\frac{2}{5} = 0.40$
3	$\frac{3}{5} = 0.60$
4	$\frac{4}{5} = 0.80$
5	$\frac{5}{5} = 1.00$

The statistician would determine the proportion of the teachers in the *sample* who do not have a teaching license (category 1) and use this as an estimate of the proportion of the *population* of primary-grade teachers who do not have a license. If 2 out of the 5 teachers in the sample do not have a teaching license, the statistician would estimate that $2/5 = 0.40$ or 40% of the population of teachers do not have a license.

The statistician could determine the chances (probability) of obtaining each of the six possible sample proportions shown in the second column of Table 13.1. These probabilities make up the *sampling distribution of a proportion.* We need not attempt to further clarify this concept, for our purposes. However, it is useful to know that the statistician bases his statistical thinking on such a sampling distribution whenever he estimates a population proportion based on a sample. Furthermore, *this sampling distribution is often well approximated by the normal distribution.* This is of the utmost importance to the statistician. By applying the sampling distribution of a proportion and by approximating this with the normal distribution, the statistician can estimate a population proportion based on sample data, make a statement as to how good this estimate is, and measure the risk that the estimate is inadequate. The sampling distribution of a proportion is also called the *binomial distribution* or the *binomial probability distribution.*

(**Note 13C** at the end of the chapter presents a more technical discussion of the sampling distribution of a proportion.)

Review

Concepts and Terminology

Binomial distribution
Binomial population
Binomial probability distribution
Normal curve
Normal distribution
Population of 1s and 0s
Sampling distribution of a proportion
Sampling distribution of the mean
Theoretical population

Problems

1. The diameters of metal fittings produced by a machine have a mean of 7.5 centimeters and a standard deviation of 0.5 centimeters. The diameters are approximately normal in distribution. What percentage of the fittings produced by the machine have diameters between—
   (a) 7.0 cm and 8.0 cm?
   (b) 6.5 cm and 7.0 cm?
   (c) 7.5 cm and 8.5 cm?
   (d) 8.0 cm and 8.5 cm?
2. Scores on a mechanical aptitude test for a population of students are approximately normal in distribution with a mean of 200 and a standard deviation of 20.

(a) Between which pair of scores are the middle 96% of the mechanical aptitude test scores?
(b) Find the score below which are 16% of the scores.
(c) Find the score below which are 84% of the scores.
(d) What percentage of the scores are above 240?

3. A large firm of lawyers who handle only divorce cases found that the number of days spent on a case during the last few years had a mean of 30 days with a standard deviation of 5 days. The firm also found that the number of days spent per case has a distribution which is approximately normal in shape. What percentage of the cases handled required—
(a) 20 to 25 days?
(b) more than 35 days?
(c) 25 to 35 days?
(d) less than 20 days?
(e) less than 40 days?
(f) more than 25 days?

4. The period of time between the acquisition of a plant virus by a certain type of leafhopper insect to transmission to plants (called the latent period) has a mean of 8 days, with a standard deviation of 2 days. The distribution of days for the latent period is approximately like the normal curve.
(a) Determine a lower limit and an upper limit for the latent period which includes the middle 68% of the distribution.
(b) Determine the latent period below which lies 16% of the distribution.
(c) Determine the latent period above which is 50% of the distribution.
(d) What percentage of the latent periods are more than 4 days in duration?
(e) What percentage of the latent periods are less than 10 days in duration?

5. What is a binomial population? Give an illustration.

6. Suppose you are interested in estimating the proportion of common stocks listed on the New York Stock Exchange that advanced in price last week. Does this imply a binomial population? Why?

7. Why is it appropriate to call a binomial population a population of 1s and 0s? Give an illustration.

8. A population of over one hundred accounts-receivable in a large firm contains about 15% delinquent accounts. What are the possible numbers and proportions of delinquent accounts in a sample of: (a) four accounts, (b) ten accounts.

9. A large inventory of tennis balls includes imported and domestic merchandise. A random sample of 20 balls is selected in order to estimate the proportion of the balls which are imports.
(a) Which is the category coded 1 and which is the category coded 0?
(b) What would you estimate as the proportion of imports in the inventory if the sample included 8 imported balls?

10. You are interested in the purchase of a large supply of peaches. However, first you wish to estimate the proportion of the peaches that are infected with brown rot. Does this imply a binomial population? Why?

11. A carload lot of apples contains some that are infected with apple scab. You are interested in the proportion of apples infected.
(a) Identify the binomial population involved.
(b) To which category would you assign the code 1 and to which the code 0? Why?

(c) What are the possible numbers and the possible proportions of infected apples in a sample of 5 selected from the carload lot?
(d) What are the possible proportions of infected apples if a sample of 20 apples is selected?

12. Compare Table 9.1, Part B (page 70) and Figure 13.1 (page 106). Based on the percentages shown for the normal distribution in Figure 13.1, explain how the following "percentages of observations included between the limits" shown in Table 9.1 (Part B), were obtained: 68, 95, 100.

13. Using Figure 13.1 or Table 9.1 (Part B), determine the limits as required.
(a) Refer to Problem 1. Determine the limits which include 95% of the metal fittings in the middle of the distribution.
(b) Refer to Problem 4. Determine the limits which include latent periods as small as one standard deviation below the mean to as much as two standard deviations above the mean. What percentage of the latent periods are included between these limits?

14. A random sample of 100 optometrists is to be selected from the practicing optometrists in a region. The sampling distribution of the mean number of hours per week worked per optometrist is approximately like the normal distribution with a mean of 40 hours per week and a standard deviation of 2 hours per week. (Note: The mean of the sampling distribution of the mean is equal to the population mean.) What is the probability the sample mean will be—
(a) within 2 hours of the population mean number of hours worked?
(b) less than 36 hours per week?
(c) more than 38 hours per week?
(d) more than 42 hours per week?
(e) no more than 4 hours per week below or above the population mean?

15. A random sample of 40 men is to be selected from a large population. If the sampling distribution of the mean weight per man is approximately normal with a mean of 180 pounds (the population mean weight) and a standard deviation of 10 pounds, what is the probability that the sample mean will be—
(a) more than 20 pounds greater than the population mean weight?
(b) more than 20 pounds below the population mean weight?
(c) within 10 pounds of the population mean weight either way?

16. A random sample of 200 retail firms is to be selected from the population of such firms in the country in order to estimate the mean volume of sales per firm during a specified week. The sampling distribution of the mean volume of sales is approximately normal with a mean of $500 (the population mean sales volume) and a standard deviation of $25. If the sample mean is used as an estimate of the population mean, what is the probability that the estimate will be—
(a) off by more than $50 either way?
(b) within $25 of the population mean either way?
(c) within $50 of the population mean either way?

17. The sampling distribution of the proportion of voters in a large city who oppose a certain tax proposal is very similar to the normal distribution for large samples. This sampling distribution for samples of 1600 voters has a mean of 0.20 and a standard deviation of 0.01. (Note: The mean of the sampling distribution of a proportion is equal to the population proportion.) If the proportion opposing the tax proposal in a sample of voters is used as an estimate of the population proportion, what is the probability that the estimate is off by more than 0.02 either way? Why?

18. In Problem 9, the following table was prepared by the statistician (he assumed that half the balls in the inventory were imports). In a sample of 20 balls:

Possible Number of Imported Balls	*Possible Proportion of Imported Balls*	*Probability of Obtaining Such a Sample*
5 or less		0.02
6		0.04
7		0.07
8		0.12
9		0.16
10		0.18
11		0.16
12		0.12
13		0.07
14		0.04
15 or more		0.02

(a) Fill in the column "possible proportion of imported balls" in the sample.
(b) What is the probability that the sample will contain from 9 to 11 imported balls?
(c) If half the balls in the inventory are imports, what is the probability that the sample proportion of imported balls will be between 0.4 and 0.6?
(d) If the sample proportion of imported balls is used as an estimate of the proportion of imports in the population (assumed to be 0.50), what is the probability that the estimate will be off by 0.15 or more either way?

19. The following table appeared in a medical journal (it was assumed that 26% of the adult males in a region suffered from some type of digestive ailment). In a sample of 25 adult males:

Possible Number with Ailment	*Possible Proportion with Ailment*	*Probability of Obtaining Such a Sample*
	At most 0.04	0.005
	0.08	0.020
	0.12	0.054
	0.16	0.104
	0.20	0.153
	0.24	0.179
	0.28	0.171
	0.32	0.135
	0.36	0.090
	0.40	0.051
	0.44	0.024
	0.48	0.010
	0.52	0.003
	0.56 or more	0.001

(a) Fill in the column of the possible number of males in the sample with the ailment.
(b) If the sample proportion of adult males with the ailment is used as an estimate of the population proportion (assumed to be 0.26), what is the probability that the estimate based on a random sample of 25 adult males is off by no more than 0.06 either way?

(c) What is the probability that the estimate is off by more than 0.10 either way?
(d) If 26% of the adult males have the ailment, what is the probability that a random sample of 25 of these males show at most 16% with the ailment?
(e) If 26% of the adult males have the ailment, what is the probability that in a random sample of 25 of these males 10 to 12 have the ailment?

20. In a certain region, 0.30 of the males 60 years of age and over have some type of heart condition. The sampling distribution of the proportion having a heart condition for samples of 1,000 men from this population has a mean of 0.30 (the same as the population proportion) and a standard deviation of 0.014. Furthermore, the sampling distribution is approximately like the normal distribution. What is the probability that a random sample of 1,000 men will include between $0.30 - 0.014 = 0.286$ (or 28.6%) and $0.30 + 0.014 = 0.314$ (or 31.4%) with a heart condition? Explain how you got your answer.

21. Use Figure 13.1 or Table 9.1 (Part B) to determine your answer.
(a) Refer to Problem 14. Determine the limits which include the central 96% of the sample means in the sampling distribution.
(b) Refer to Problem 17. Determine the limits which include the central 68% of the sample proportions in the sampling distribution.
(c) Refer to Problem 17. Determine the range of sample proportions which include all (actually, practically all) of the possible sample proportions of voters who oppose the tax proposal.

Note 13A

Normal Distribution and Applications

The normal distribution is a continuous distribution of an infinite population. That is, the variable of the distribution X_i, or more simply X, is continuous and can take on any value (from negative infinity to positive infinity). The variable on the horizontal axis of the normal curve shown in the bottom part of Figure 13.1 is X. Theoretically, this axis should stretch out indefinitely in both directions and the lower (left) and upper (right) tails of the distribution should extend out indefinitely, getting closer to the horizontal axis, but never touching it.

There are many normal distributions, and a particular distribution is identified when the mean and standard deviation are specified. Figure 13A.1 shows that when the mean of a normal

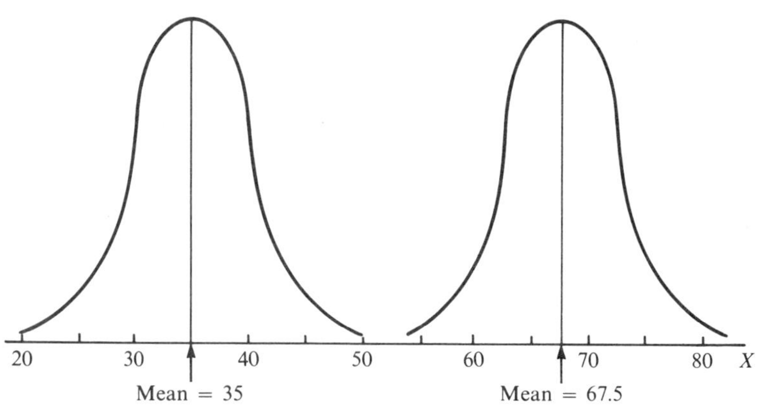

Figure 13A.1

Normal Distributions with the Same Standard Deviation but Different Means

distribution is changed (from 35 to 67.5) and the standard deviation remains the same, the only effect is to shift the distribution along the horizontal axis.

However, if the standard deviation is changed, the effect is to change the appearance of the curve. Notice in Figure 13A.2 that the two distributions have the same mean (60), but the distribution with a standard deviation of 5 shows a contraction of the central portion of the distribution and a higher peak compared to the curve with a standard deviation of 10. Clearly, the greater the standard deviation, the more spread out is the distribution, as is to be expected from the meaning of the standard deviation (discussed in chapter 9).

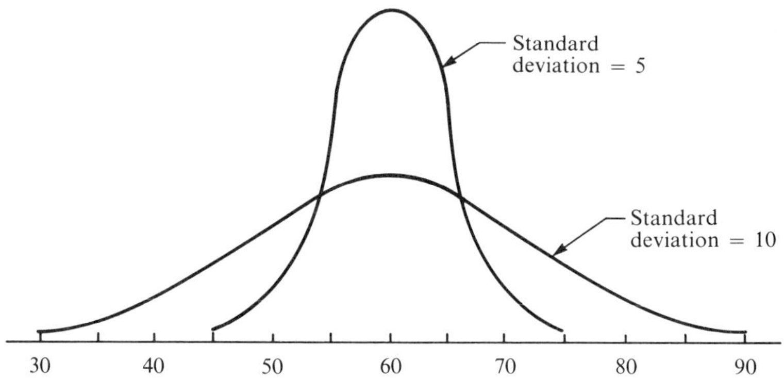

Figure 13A.2

Normal Distributions with the Same Mean and Different Standard Deviations

Generally, statisticians deal with the *standard normal distribution* where the variable X is expressed in standard units which we denote as z. We discussed standard units in chapter 9. The equations to convert a variable X to standard units z are

For population data:

$$z = \frac{X - \mu}{\sigma}$$

For sample data:

$$z = \frac{X - \bar{X}}{s}$$

where X = a specified value of a variable
μ = population mean (μ is the Greek letter "mu")
σ = population standard deviation (σ is the Greek letter "sigma")
$\bar{X}$ = sample mean (read "X bar")
s = sample standard deviation

When expressed in standard units, all normal distributions have the same appearence and they all have a mean equal to 0 and a standard deviation equal to 1. Figure 13A.3 presents the standard normal distribution (notice its similarity to Figure 13.1), where the variable on the horizontal scale is z (not X) and the mean is at $z = 0$. As discussed in chapter 9, standard units (z values) are in standard deviation units; so $z = 1$ means one standard deviation above the mean, $z = -2$ means two standard deviations below the mean, as indicated in Figure 13A.3.

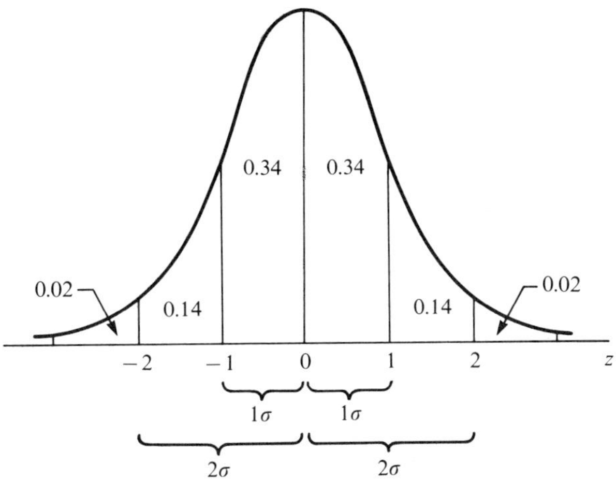

Figure 13A.3

The Standard Normal Distribution (Mean = 0, Standard Deviation = 1)

Statisticians base most of their statistical inferences on the proportion of the area under the normal curve for specified intervals along the horizontal axis. Table B.1 in Appendix B is useful for such purposes. This table is based on the standard normal curve and presents areas under the curve. The shaded area in the normal curve at the top of Table B.1 indicates that the table shows the proportion of the total area under the curve between the mean and a specified z value. The total area under the standard normal curve is equal to 1.00.

Let us determine the area between the mean and $z = 1.00$. Look down the column headed "z" until you reach 1.0; then move across that row to the column headed "0.00." The value 0.3413 obtained, corresponding to $z = 1.00$, is the proportion of the area under the curve between the mean and one standard deviation above the mean. In other words, 34.13% of the total area under the normal curve is between the mean and $z = 1.00$. This agrees with 0.34 shown in Figure 13A.3 after rounding to two decimals.

Table B.1 shows areas only for positive z values. If $z = -1.96$, we obtain the area under the curve between the mean and this z value as follows: Find 1.9 in the z column, then move across this row to the column headed "0.06." The value 0.4750 obtained corresponds to $z = 1.96$ (the area between the mean and $z = 1.96$). However, because of the symmetry of the normal distribution, 0.4750 is also the area between the mean and $z = -1.96$.

What is the area under the normal curve between $z = -1.50$ and $z = 1.39$? As shown in Table B.1, the area between $z = -1.50$ and the mean is 0.4332. The area between the mean and $z = 1.39$ is 0.4177. Therefore, we obtain the area between these two z values by adding 0.4332 + 0.4177 to obtain 0.8509, as shown in Figure 13A.4.

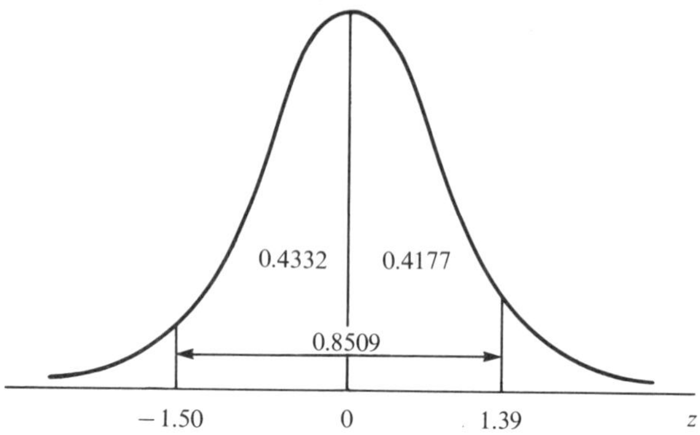

Figure 13A.4

Determining the Area under the Normal Curve between a Negative z Value (−1.50) and a Positive z Value (1.39)

Suppose a population of weights is approximately normal in distribution with a mean of 224.83 pounds and a standard deviation of 39.50 pounds. Determine the proportion of the weights between 185.00 pounds and 210.00 pounds. First transform these weights to standard units as follows:

$$z_1 = \frac{185.00 - 224.83}{39.50} = -1.01$$

$$z_2 = \frac{210.00 - 224.83}{39.50} = -0.38$$

Referring to Table B.1, the area corresponding to z_1 is 0.3438 and the area corresponding to z_2 is 0.1480. Then, subtracting 0.3438 − 0.1480, we obtain 0.1958 as the proportion of weights in the population between the specified limits. This is illustrated in Figure 13A.5.

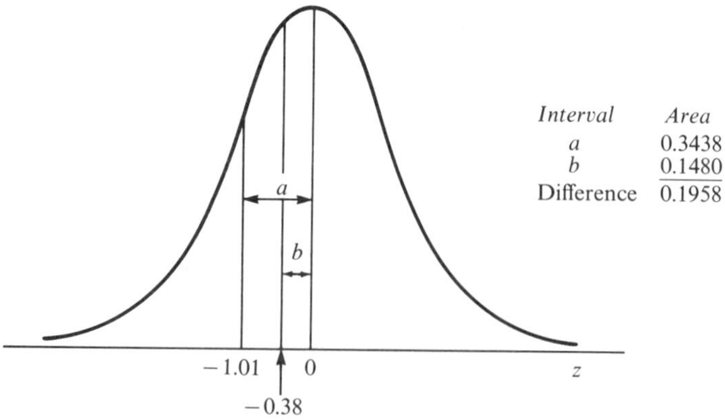

Figure 13A.5

Determining the Area under the Normal Curve between Two Negative z Values

Since the area under the normal curve, corresponding to an interval along the horizontal axis, represents a proportion, it may also be interpreted as a probability. This is the principal use of the normal distribution in statistics. In the foregoing illustration, if a weight is selected at random from the population of weights, the probability is 0.1958 (or about 0.20) that it will be between 185.00 pounds and 210.00 pounds.

Suppose a large population of mutual fund investors added a mean of \$118.35 to their investments last month, with a standard deviation of \$22.10. Assuming that the investment amounts are approximately normal in distribution, find the dollar value below which 10% of the investments last month fell. Figure 13A.6 presents the standard normal curve for the distribution of investment amounts. The area in the lower tail represents the lower 10%, or 0.1000, of the investment amounts last month. Therefore, since the normal distribution is symmetrical and 50%, or 0.5000, of the total area under the curve is below the mean, the area below the mean not in the lower tail is 0.5000 − 0.1000, or 0.4000.

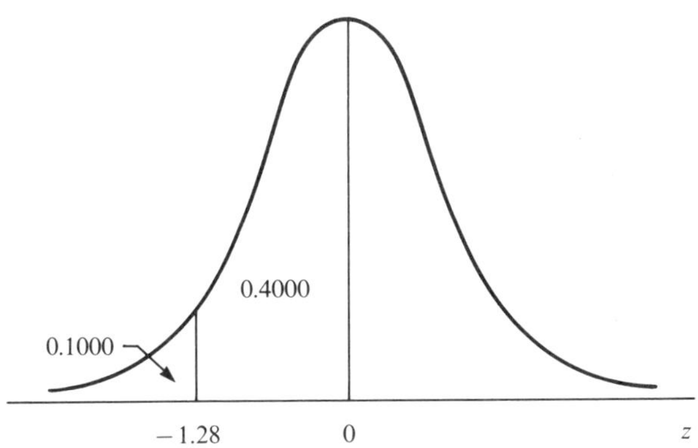

Figure 13A.6

Determining the z Value Corresponding to a Specified Area under the Normal Curve (10% in the Lower Tail)

Refer to Table B.1 and find 0.4000 (or a value as close to this as possible) in the body of the table. The closest value is 0.3997, which corresponds to $z = 1.28$. Then, $z = -1.28$ marks off about 0.4000 (actually 0.3997) of the area under the curve between the mean and this z value, and about 0.1000 (or 10%) of the area lies in the lower tail. This is shown in Figure 13A.6. Therefore, $z = -1.28$ is the investment amount we want expressed in standard units. Substituting in the equation for z we have the following:

$$-1.28 = \frac{X - 118.35}{22.10}$$

Solving for X, we obtain $X = \$90.06$. Then, 10% of the investments last month were less than \$90.06.

Note 13B

Sampling Distribution of the Mean

Let μ = population mean
σ = population standard deviation
$\bar{X}$ = sample mean (read "X bar")

If a large population has a mean μ and a standard deviation σ and samples of size n (= 30 or more) are selected, then the sampling distribution of the mean may be approximated by the normal distribution. The mean of this sampling distribution is equal to the population mean, and the standard deviation is equal to $\sigma/\sqrt{n}$.

The standard deviation of a sampling distribution is called the *standard error*. Notice that no matter how the population is distributed (with little restriction), the sampling distribution of the mean is approximately like the normal curve, as long as the population is large and the sample size is 30 or more. (Actually, if the population distribution itself is approximately normal, the sample size could be less than 30.)

What is the probability that a sample of 36 lengths randomly selected from a large population of lengths (with μ = 160 centimeters and σ = 18 centimeters) has a mean length greater than 165 centimeters?

A sample mean $\bar{X}$ is transformed to standard units by computing

$$z = \frac{\bar{X} - \mu}{\frac{\sigma}{\sqrt{n}}}$$

Transforming 165 centimeters to standard units, we obtain the following:

$$z = \frac{165 - 160}{\frac{18}{\sqrt{36}}} = 1.67$$

Then, referring to Table B.1 in Appendix B, we determine that the area under the normal curve corresponding to $z = 1.67$ is 0.4525. Hence, the required probability is computed as $0.5000 - 0.4525 = 0.0475$ (the area above 165 centimeters in the normal distribution of lengths). In other words, the chances are about 48 in 1,000 that the mean of a random sample of 36 weights will be more than 165 centimeters.

Notice that the *standard error of the mean* $\sigma/\sqrt{n}$ is larger for larger σ and smaller for larger n. This standard error measures the variation of sample means around the population mean, so the smaller this standard error, the closer are possible sample means to μ and the more likely is it that $\bar{X}$ will provide a good estimate of μ.

Note 13C

Sampling Distribution of a Proportion

Let p = proportion of items in a population with a specified characteristic (or proportion of 1s in a binomial population)

X = number of observations in a sample with the specified characteristic (or number of 1s in the sample)

n = sample size

$\frac{X}{n}$ = proportion of observations in the sample with the specified characteristic (or proportion of 1s in the sample)

If a large binomial population has the proportion p and samples of size n (sufficiently large) are selected, then the sampling distribution of a proportion may be approximated by the normal distribution. The mean of this sampling distribution is equal to the population proportion, and the standard deviation (called the *standard error of a proportion*) is equal to $\sqrt{p \times (1 - p)/n}$.

The foregoing description of the sampling distribution of a proportion applies only if the population is large, p is not close to 0 or 1, and the sample size is at least 100. (A larger sample is usually preferred).

Suppose a ton of seed includes 20% of a special type. What is the probability that a random sample of 500 seeds will include no more than 15% of the special type of seed?

Since the sampling distribution of a proportion may be approximated by the normal distribution, we may use Table B.1 in Appendix B to determine the required probability. The sample proportion is transformed to standard units as follows:

$$z = \frac{\frac{X}{n} - p}{\sqrt{\frac{p \times (1 - p)}{n}}}$$

The standard error of a proportion for our problem is computed (with $n = 500$ and $p = 0.20$) as follows:

$$\sqrt{\frac{0.20 \times (1 - 0.20)}{500}} = 0.018$$

Then, we transform the specified sample proportion $X/n = 0.15$ to standard units as follows:

$$z = \frac{0.15 - 0.20}{0.018} = -2.78$$

Referring to Table B.1, the area corresponding to $z = -2.78$ is 0.4973. Then, the probability that 15% or less of the sample will be the special type of seed is computed as $0.5000 - 0.4973 = 0.0027$ (or about 3 chances in 1,000).

Part D

POPULATION ESTIMATES AND DECISION MAKING

14

Sample Surveys and Population Estimates

Sample surveys are used in all fields of business and professional activity. Sometimes these surveys are large-scale market studies, sometimes they are relatively small sampling projects. Their objective is estimation of one or more population characteristics. For example, a sample of 300 cities with populations of 25,000 or less is selected to estimate the *average* (*mean*) expenditure for education per pupil for all such cities in the nation. Or, a sample of 75 third-graders is selected to try out a new reading-progress test in order to estimate the *proportion* of all third-graders in a city who would be expected to obtain a grade below a specified level on the test. A department store needs to estimate the *total dollar amount* of unpaid balances and selects a sample of delinquent customer-accounts for this purpose. Or, a social caseworker wishes to estimate the *total number* of welfare recipients who found employment but did not report it and selects a sample of 100 welfare recipients in order to prepare this estimate.

We have just illustrated the use of samples to estimate a population *mean*, the *total amount* in a population, a population *proportion*, and the *total number* in a population category. These represent the varieties of estimates frequently encountered.

14.1 Statistical Estimation and Confidence Intervals

The statistician conceives of two types of estimates, *point estimates* and *interval estimates*. Usually, he determines both types. As an illustration, suppose it is required to estimate the average (mean) expenditure per family for health care during the past month in a large city. Assume the statistician on this project selects a random sample of 250 families in the city and computes the mean expenditure per family in the sample as \$25. Based on this sample mean, he estimates that mean family expenditure for health care for all families in the city during the past month was \$25.

In other words, the statistician uses the sample mean as an estimate of the population mean. This is a point estimate. A *point estimate* is a single value estimate and the sample mean is used as a point estimate of the population mean.

Of course, it is not likely that a point estimate will be exactly correct. As we have seen in Tables 12.2 and 12.4, many different sample means (and therefore many different point estimates) are possible based on random sampling from a population. It is customary for the statistician to accompany a point estimate with some kind of measure to indicate *how good* an estimate it is.

For example, he might state that, based on the sample, he estimates that the population mean expenditure per family was \$25 ± \$5 (read as "\$25 + or − \$5"). In other words, the population mean expenditure could be as low as 25 − 5 = \$20 or as high as 25 + 5 = \$30 per family.

Or, the statistician could state that he estimates the population mean expenditure as \$25 per family, but the mean could be anywhere in the interval \$20 to \$30. Of course, an *interval estimate* (such as \$20–\$30) provides a safer estimate of a population mean than a point estimate. However, an interval estimate is not useful unless it is accompanied by a statement of *how confident* we can be that the interval includes the population mean. For example, the statistician might state that he is 95% confident that the interval \$20–\$30 includes the population mean.

Similarly, statisticians prepare interval as well as point estimates for population proportions, total amounts, and total numbers and accompany each interval estimate with a statement of how much confidence we may have that the stated interval includes the true population value. Such an interval estimate is called a *confidence interval.* A *confidence interval* indicates how reliable (how good) the point estimate is.

14.2 Estimating the Population Mean and Total Amount

Before a sample is selected and population estimates are prepared, the user of the estimates must make certain decisions. We know that an estimate based on sample data is not likely to be exactly correct. Consequently, the user must decide *how good* an estimate is required. "How good" means how much error is acceptable and how much confidence is required that the error does not exceed the amount specified. The statistician must take these user specifications into account when deciding on the size of sample to select and the way to prepare his estimates. If these specifications are not met, the estimates will not be adequate for the user's purposes.

Suppose it is desired to estimate both the average (mean) acreage per farm and the total acreage of the 10,000 farms in a region. Assume a sample of 200 farms is selected and the mean number of acres per farm in the sample is computed as 210 acres. (The statistician will also need to compute the sample standard deviation.) The statistician could present his estimate of the mean acreage per farm in the population in one of the following two ways:

(a) Point estimate: 210 acres
95% (or 0.95) confidence interval : 200–220 acres

(b) 210 ± 10 acres, at the 95% (or 0.95) level of confidence

Let us introduce some new terminology. The interval 200–220 acres (the *confidence interval*) is determined by the *confidence limits*, 200 (*lower confidence limit*) and 220 (*upper confidence limit*). The confidence specification 95% (or 0.95) is called the *confidence coefficient*.

How is a *confidence interval* for a population mean determined? Generally speaking, it is determined on the basis of the mean and standard deviation computed from a random sample, the sample size, a sampling distribution similar to those in Tables 12.2 and 12.4, and the use of the normal distribution to approximate this sampling distribution. (The sampling distribution involved is called the *sampling distribution of the mean*.) (**Note 14A** at the end of this chapter discusses how to construct confidence intervals for a population mean.)

Let us examine what a confidence interval means. Imagine that the sample of 200 farms is returned to the population, another sample of 200 farms is selected, and a confidence interval is computed in the same way as the first one. Then, this sample of farms is returned to the population, another sample of 200 farms is selected, and (again) a confidence interval is computed. Imagine that this *repeated sampling procedure* is carried out a large number of times and that each time a confidence interval is computed. This will result in the computation of a large number of confidence intervals. It has been shown mathematically that if the same procedure used to construct the 95% confidence interval in our illustration (200–220 acres) is used each time, then 95% of the confidence intervals computed will actually include the population mean. The other 5% will not.

Consequently, the statistician reasons that the computational procedure used to construct a confidence interval, when repeatedly applied to random samples from a population, results in intervals that include the population mean 95 times in 100. Hence, the probability is 0.95 that the confidence interval constructed for a *particular sample* will include the population mean. Furthermore, such a high probability gives a high feeling of confidence that a particular interval computed will actually include the population mean. Such an interval is called a *95% confidence interval* or is said to be constructed at the *95% level of confidence*. Finally, the computation 100% − 95% = 5% indicates that there is a 5% risk (5 chances in 100) that the particular confidence interval constructed will not include the population mean.

Notice that in the first estimation statement, the statistician displays the point estimate (210 acres) and the confidence interval (200–220 acres). Subtracting 210 (point estimate) − 200 (lower confidence limit), we obtain 10 acres. Subtracting 220 (upper confidence limit) − 210 (point estimate), we also obtain 10 acres. Then, if 210 acres is used as an estimate of the population mean, we have 95% confidence (confidence coefficient) that the estimate is off by no more than 10 acres either way. In other words, we are 95% confident that the *maximum error* we may make in using this estimate is 10 acres. Consequently, we are taking a 5% risk that the maximum error could be greater than 10 acres and that the estimate is inadequate.

In the second estimation statement, the statistician displays the point estimate (210 acres) and the maximum error expected (10 acres). This form of presentation is often preferred. Then, the confidence limits may easily be determined as follows: 210 − 10 = 200 acres (lower confidence limit) and 210 + 10 = 220 acres (upper confidence limit).

The confidence interval we have been discussing is the *confidence interval for the mean*. It may be desired to compute an estimate of the total acreage of the 10,000 farms in the region (in our illustrative problem) and a 95% confidence interval for this estimate. Such an interval estimate is a *confidence interval for a total amount*. Since there are 10,000 farms in the population, we may determine a point estimate of total acreage (total amount) as 10,000 × 210 (point estimate of the mean) = 2,100,000 acres. Then, 10,000 × 10 (maximum error expected for the estimate of the mean) = 100,000 acres represents the maximum error expected in the estimated total acreage (at the 95% level of confidence). The estimation statement for total acreage of the 10,000 farms in the region would be: 2,100,000 ± 100,000 acres, at the 95% level of confidence.

The confidence limits may be computed as 2,100,000 − 100,000 = 2,000,000 acres (lower confidence limit) and 2,100,000 + 100,000 = 2,200,000 acres (upper confidence limit). Then, the 95% confidence interval is 2,000,000–2,200,000 acres. This indicates that there is a risk of 5 in 100 that the estimated total acreage of 2,100,000 is off by more than 100,000 acres (either way) and that total acreage is either less than 2,000,000 or more than 2,200,000.

14.3 Estimating the Population Proportion and Total Number

Very often the objective of a sample survey is to estimate a population *proportion* or the *total number* in a population that falls in a specified category. For example, a political analyst may conduct a sample survey in a city to estimate the proportion of the voters who favor a certain candidate running for city mayor. Or, an insurance firm may engage a statistical consultant to estimate the total number of families in a city who do not have health insurance.

Suppose the statistical consultant selects a sample of 500 families from the 50,000 families in the city and it is desired to estimate both the proportion and total number of families who do not carry health insurance. He determines that 9% of the families in the sample do not carry health insurance. Then he prepares his estimate of the population proportion as follows:

0.09 ± 0.025, at the 95% level of confidence

Notice that the *sample proportion* (0.09 or 9%) is used as an estimate of the *population proportion*. The *maximum error expected* either way when using this estimate is 0.025, at the 95% level of confidence (as shown in the foregoing estimation statement). The lower confidence limit is computed as 0.09 − 0.025 = 0.065 or 6.5%, and the upper confidence limit is computed as 0.09 + 0.025 = 0.115 or 11.5%. Then, the 95% confidence interval is 0.065–0.115. It may also be written as 6.5%–11.5%. In other words, we are 95% confident that the percentage of families in the city who do not carry health insurance is between 6.5% and 11.5%.

The *confidence interval for a proportion* is based on the concept of repeated sampling from a population (as previously explained for the mean). When estimating a proportion, the statistician thinks of the population of families as being a *binomial population* (families with health insurance and families with no health insurance). The confidence interval is computed based on the sample proportion in the category of interest (families with no health insurance), the sample size, the sampling distribution of a proportion, and the use of the normal distribution to approximate this sampling distribution. (**Note 14B** , at the end of the chapter, discusses how to construct confidence intervals for a population proportion.)

The foregoing estimation statement indicates that there is a 5% risk that the estimated population proportion 0.09 is off by more than 0.025 either way and that this proportion may be below 0.065 or above 0.115.

Since there are 50,000 families in the city, the statistical consultant computes a point estimate of the *total number* of families with no health insurance as 50,000 × 0.09 (point estimate of the population proportion) = 4,500. Then, 50,000 × 0.025 (maximum error expected for the population proportion) = 1,250 represents the maximum error expected for the estimated total number of families with no health insurance (at the 95% confidence level). The following estimation statement is presented for the total number of families with no health insurance in the city:

4,500 ± 1,250 families, at the 95% level of confidence

The confidence limits are computed as 4,500 − 1,250 = 3,250 families (lower confidence limit) and 4,500 + 1,250 = 5,750 families (upper confidence limit). Then, the 95% confidence interval is 3,250–5,750 families. This indicates that there is a risk of 5 chances in 100 that the estimated total number of families with no health insurance in the city (4,500) is off by more than 1,250 families and that the number of families in this category is either less than 3,250 or more than 5,750.

14.4 How Good an Estimate Do You Need ?

We have been discussing 95% confidence intervals. However, confidence intervals can be constructed at any desired level of confidence. Alternative levels of confidence frequently used are the 98% level and the 99% level. Say a sample of 150 apartments is selected in a metropolitan area to estimate the average (mean) rental of all apartments in the area, and the mean rental in the sample is $190. The following estimation statements might be computed by the statistician:

(a) $190 ± $29.40, at the 95% level of confidence
(confidence interval: $160.60–$219.40)

(b) $190 ± $34.95, at the 98% level of confidence
(confidence interval: $155.05–$224.95)

(c) $190 ± $38.70, at the 99% level of confidence
(confidence interval: $151.30–$228.70)

The first estimation statement indicates that we can be 95% confident that the estimate of \$190 as the mean rental for apartments in the metropolitan area is off by no more than \$29.40 either way. The second estimation statement indicates that we can be *more confident* (98% confident) that the estimate is off by no more than \$34.95 either way. The third estimation statement indicates that we can be *almost certain* (99% confident) that the estimate is off no more than \$38.70 either way. Clearly, the more confident we can be of the maximum error expected, the larger is the expected maximum error.

Alternatively, we can compare the foregoing estimation statements from the point of view of the confidence intervals. The first statement indicates that we can be 95% confident that the interval \$160.60–\$219.40 includes the mean rental for the population (all apartments in the metropolitan area). Subtracting \$219.40 − \$160.60, we obtain \$58.80, which is the *width* of the confidence interval. The second statement indicates that we can be 98% confident that the interval \$155.05–\$224.95 includes the population mean rental. This interval is of width \$69.90. The third statement indicates that we can be nearly certain (99% confident) that the interval \$151.30–\$228.70 includes the population mean rental, and the width of this interval is \$77.40. Notice that the more confident we can be that a stated interval includes the mean, *the wider is the interval.* (The width of a confidence interval is equal to twice the maximum error expected.)

The narrower the confidence interval and the higher the level of confidence, the better is the estimate. For example, the confidence interval for the mean 200 ± 5 is preferable to the interval 200 ± 10, if both are at the same level of confidence. Clearly, an expected maximum error of 5 (indicating a confidence interval width of 10) indicates a better estimate than an expected maximum error of 10 (indicating a width of 20). Also, the confidence interval 200 ± 5 at the 99% confidence level is preferable to the same interval at the 95% level because it indicates a higher confidence that the interval includes the population mean. Also, it carries a smaller risk that the estimate is inadequate (1% for a 99% interval compared with a 5% risk for a 95% interval).

It would seem to be in the user's best interests for the statistician always to prepare his estimates with a very small expected maximum error (so that the confidence interval will be narrow) and to provide a high level of confidence. Actually, he can always do this if he uses a large enough sample. However, the larger the sample, the more costly the survey and the longer it will take to produce the desired estimates.

Consequently, the first step in planning a sample survey for the purpose of estimating one or more population characteristics is for the user to determine *how good* an estimate he needs. More specifically, the user must decide (with the assistance of the statistician, if necessary) *how much error* he is willing to accept and *how much assurance* (*confidence*) he requires that the maximum error expected in the estimate is no greater than specified. Sometimes, a rather large error and a 95% confidence level is quite sufficient.

For example, the real estate assessment department in a city may request the statistician to estimate the mean market value of single-family homes in the city and specify that only a fair estimate is quite adequate for the needs of the department. Of course, they will have to state their requirements more specifically. They may decide, perhaps with the aid of the statistician, that an estimate of mean market value that may be off by as much as $2,500 either way (but no more) is adequate and that they would be satisfied with 95% assurance that the maximum error is no greater than specified. Then, it would be unnecessary for the statistician to prepare a better (and more costly) estimate.

On the other hand, a prospective purchaser of a large lot of copper fittings may need a very good estimate of the proportion of fittings in the lot that are defective, because he will use this estimate to make a bid on the purchase of the entire lot of fittings. He may state that he requires an estimate of the proportion of defective fittings in the lot that is off by no more than one percentage point either way and he wants to be 99% confident that the estimate is not off by any greater amount. Such an estimate may require a very large sample.

The user should carefully consider his requirements. Requesting estimates better than needed could lead to unnecessary expense and delays in carrying out the survey and preparing the estimates. (**Note 14C,** at the end of the chapter, shows how to determine the sample size required in a given situation.)

Review

Concepts and Terminology

95% (or 0.95) confidence interval
98% (or 0.98) confidence interval
99% (or 0.99) confidence interval
Confidence coefficient
Confidence interval
Confidence interval for a proportion
Confidence interval for a total amount
Confidence interval for a total number
Confidence interval for the mean
Confidence limits
Level of confidence
Lower confidence limit
Maximum error expected
Point estimate
Reliability of an estimate
Sample surveys
Upper confidence limit
Width of a confidence interval

Problems

1. Based on sample data, how does the statistician estimate (a) the population mean, (b) a population proportion?
2. Why is an interval estimate "safer" than a point estimate? Give an illustration.
3. What is meant when we speak of "how good" an estimate is?
4. Based on the concept of repeated sampling from a population, how would you explain the meaning of (a) 95% confidence interval, (b) 98% confidence interval, (c) 99% confidence interval?

5. A sample of 50 mortgage recordings was randomly selected from a population of 5,000 home mortgage recordings in order to estimate the population mean amount of recordings. The statistician presents his estimate as follows:

$$\$30{,}000 \pm \$2{,}000 \text{ (at the .95 level of confidence)}$$

(a) What is the confidence coefficient?
(b) State the confidence interval for the population mean.
(c) What is the maximum error expected at the 95% level of confidence if $30,000 is used as an estimate of the population mean?
(d) What is the width of the confidence interval?
(e) What is the point estimate of the population mean amount of recordings?

6. What is the probability that the procedure used in constructing the confidence interval in Problem 5 results in an interval that does not include the population mean?

7. In Problem 5—
(a) Compute a point estimate of the total dollar amount of home mortgage recordings in the population.
(b) What is the maximum error you could expect to make in using this point estimate, at the 0.95 level of confidence?
(c) Determine a 0.95 confidence interval for the total dollar amount of home mortgage recordings.

8. A sample of 60 cases was randomly selected from the records of a circuit court for the past three years in order to estimate the mean number of court days spent per case. Based on the sample data the statistician estimated the population mean number of days as 10 ± 3, at the 0.98 level of confidence.
(a) What is the confidence coefficient?
(b) Determine the 98% confidence interval for the population mean.
(c) What is the width of the confidence interval?
(d) What is the point estimate of the population mean?
(e) What is the maximum error you could expect to make at the 0.98 confidence level if this point estimate is used?
(f) What is the probability that the procedure used to compute the confidence interval produces an interval that includes the population mean? Does not include the population mean?

9. In Problem 8, assume that 800 cases were on the court's records for the three-year period.
(a) Compute a point estimate of the total number of court days spent on these cases.
(b) What is the maximum error you could expect to make, at the 0.98 confidence level, if you use this estimate?
(c) Construct a 0.98 confidence interval for the total number of court days.

10. A random sample of 40 small cities showed a mean of 10 traffic accidents last month. The statistician indicated that if the sample mean is used as an estimate of the mean for the population of small cities he could expect to be off by no more than 2 accidents at the 95% level confidence.
(a) Construct a 0.95 confidence interval for the population mean.
(b) What is the width of this confidence interval?
(c) What is the confidence coefficient of this interval estimate?
(d) What is the point estimate of the population mean?

11. In Problem 10, there are 500 cities in the population of cities.
(a) Determine a point estimate of the total number of accidents in the population.

(b) Construct a 0.95 confidence interval for the total number of accidents in the population.

12. A random sample of 75 savings and loan associations in a region showed a mean of 30 mortgage loans written last month. According to the statistician, if the sample mean is used as an estimate of the mean number of mortgage loans written last month in all the saving and loan associations in the region, the maximum error to be expected is 5 loans at the 0.95 level of confidence and 6 loans at the 0.98 level of confidence.

(a) Construct a 95% confidence interval for the population mean.
(b) Construct a 98% confidence interval for the population mean.
(c) If a 0.99 confidence interval were constructed, how would it compare with the intervals constructed at the 0.95 and 0.98 levels of confidence? Defend your answer.

13. In Problem 12, it was determined that there are 400 saving and loan associations in the region. Construct a confidence interval for the total number of mortgage loans written by these associations last month at the (a) 0.95 level of confidence, (b) 0.98 level of confidence.

14. A sample of 100 families randomly selected from 10,000 families in a city included 64 with at least one teenager in the family. The statistician presents his estimate of the proportion of families in the population with at least one teenager in the family as follows:

$$0.64 \pm 0.112, \text{ at the } 98\% \text{ level of confidence}$$

(a) What is the confidence coefficient?
(b) Determine the 0.98 confidence interval for the population proportion.
(c) What is the maximum error expected at the 0.98 confidence level if 0.64 is used as an estimate of the population proportion?
(d) What is the width of the confidence interval?
(e) If the confidence interval were redetermined at the 0.95 level, would the confidence interval be narrower, wider, or the same as for the 0.98 level? How would the maximum error expected be affected?

15. What is the probability that the method used to construct the confidence interval in Problem 14 results in an interval that does not include the population proportion?

16. In Problem 14—

(a) Determine a point estimate of the number of families in the population with at least one teenager in a family.
(b) What is the maximum error you would expect to make at the 98% level of confidence if this point estimate is used?
(c) Determine a 98% confidence interval for the number of families in the population with at least one teenager in the family.

17. A sample of 200 plants of a certain variety randomly selected from a large supply includes 60 that show some wilting. The statistician presents his estimate of the proportion of the full supply of plants that are affected by wilting as 0.30 ± 0.063, at the 95% confidence level.

(a) What is the point estimate of the proportion of plants with some wilt in the population? How was it obtained?
(b) Determine the lower and upper limits of the 0.95 confidence interval.
(c) What is the probability that the procedure involved in constructing the confidence interval above would produce an interval that includes the true population proportion? Does not include this proportion?

18. In Problem 17, it was determined that 3,000 plants make up the total supply.

(a) Determine a point estimate of the number of plants in the total supply that are affected by wilting.

(b) What is the maximum error you would expect to make at the 0.95 level of confidence if the point estimate is used?

(c) Construct a 0.95 confidence interval for the number of plants in the total supply that are affected by wilting.

19. A sample of 60 peaches randomly selected from a large supply shows that 15 are infected with brown rot.

(a) Compute a point estimate of the proportion of peaches in the total supply infected with brown rot.

(b) It is determined that the maximum error expected at the 0.99 confidence level is 0.143 if this point estimate is used. Construct a 0.99 confidence interval for the population proportion.

(c) It is determined that the maximum error expected is 0.130, at the 0.98 confidence level. Construct a 0.98 confidence interval.

(d) A 95% confidence interval was constructed as follows: 0.14–0.36. What is the maximum error expected at the 95% level of confidence?

(e) What is the probability that a confidence interval for the population proportion will include this proportion, if determined according to the procedure used for (b), (c), and (d), above?

(f) What is the probability the procedures used for (b), (c), and (d), above, produce confidence intervals that do not include the population proportion?

20. In Problem 19, 5,000 peaches make up the supply.

(a) Compute a point estimate of the number of peaches in the full supply that are infected with brown rot.

(b) Construct a 0.98 confidence interval for the number of infected peaches in the supply.

(c) Construct a 0.95 confidence interval for the total number of infected peaches.

21. A computer software firm is considering the development of a software package specifically for time-sharing computer use. The manager asks his statistician to conduct a survey of business firms in a specified region to determine the number of firms that use time-sharing computer facilities. He specifies that he is interested in a rough estimate (an expected error of 50 firms at the 0.95 confidence level is quite satisfactory). The statistician determines the following estimate: 1,800 ± 10, 0.95 confidence level. Comment on the usefulness and cost characteristics of this estimate.

22. A health-study team engaged a statistician to conduct a survey of the high-school student population in a city to determine the average number of dental visits made by a student during the last year. The statistician was advised that an estimate that could be off by as many as 5 visits, with a confidence level of 0.95, was satisfactory. The statistician selected a random sample of students and constructed the following estimate:

(a) 15 ± 2 visits, at the 0.98 level of confidence. Does this result satisfy the estimation requirements imposed on the statistician? Defend your answer.

(b) 15 ± 10 visits, at the 0.95 confidence level. Does this result satisfy the estimation requirements? Defend your answer.

(c) 15 ± 4 visits, at the 0.95 confidence level. Is this confidence interval satisfactory? Why?

Note 14A

Constructing Confidence Intervals for a Population Mean

This note presents a method for constructing a confidence interval for a population mean where (1) the population is large, and (2) a sample of size 30 or more is selected.

Note 13B discussed and described the sampling distribution of the mean. This sampling distribution is illustrated by the normal curve in Figure 14A.1(a). Any $\bar{X}$ value (sample mean) along the horizontal scale of the chart represents a possible sample mean. When a sample is selected and $\bar{X}$ is computed and used as a point estimate of the population mean, we do not know how good an estimate we have. Fortunately, most of the $\bar{X}$ values possible are concentrated around the population mean μ, as indicated by the central peak in the sampling distribution. Hence, the chances are quite good that a particular $\bar{X}$ provides an acceptable estimate of the population mean.

$\bar{X}_a$ and $\bar{X}_b$, shown in Figure 14A.1(a), are two possible sample means that correspond to $z = -1.96$ and $z = 1.96$. As can be determined by the use of Table B.1 in Appendix B, these are the limits between which lie 0.95 of the possible sample means. Alternatively, the probability is 0.95 that the mean of a random sample will lie in the interval defined by these limits.

In order to construct a confidence interval based on a sample, we must first determine the sample size (n), the sample mean ($\bar{X}$), and the sample standard deviation (s). (**Note 9A** shows how to compute s). When using $\bar{X}$ as a point estimate of μ, we compute the maximum expected error (m) at a stated confidence level as follows:

$$m = z \times \frac{s}{\sqrt{n}} \times \sqrt{\frac{N - n}{N - 1}}$$

The value of z depends upon the confidence coefficient in a given problem. For a 0.95 confidence interval, $z = 1.96$; so 95% of the area under the normal curve is included between $z = -1.96$

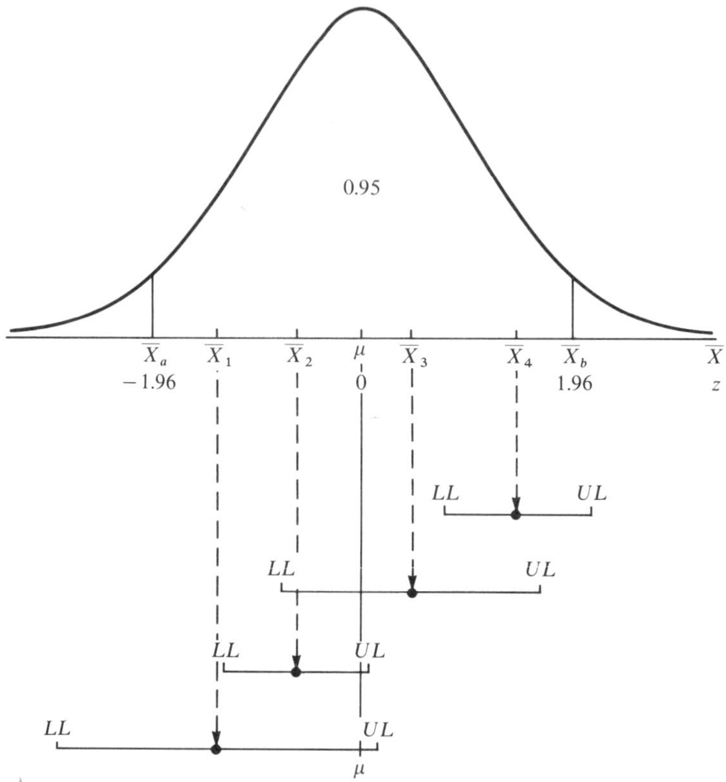

Figure 14A.1

(a) Sampling Distribution of the Mean, and (b) Illustrative Confidence Intervals

and $z = 1.96$ (as shown in Figure 14A.1(a)). For a 0.98 confidence interval, $z = 2.33$; so 98% of the area under the curve is included between $z = -2.33$ and $z = 2.33$ (as may be determined by the use of Table B.1). The factor $s/\sqrt{n}$ in the foregoing equation is the *estimated standard error of the mean.*

The factor $\sqrt{(N - n)/(N - 1)}$ in the equation for m is called the *finite population correction* factor (abbreviated as fpc), where N denotes population *size*. The statistical theory underlying the equation for m (with the fpc ignored) is based on sampling from an infinite population (or sampling with replacement from a finite population). When sampling without replacement from a finite population (the usual situation in practice), the fpc must be included in the equation for m in order to make a proper adjustment because sampling is without replacement. If the sample size n is negligible compared to the population size N, the fpc is so close to 1.00 that it may be ignored. Statisticians usually ignore the fpc when the *sampling ratio* n/N is less than 0.05.

Finally, a confidence interval is constructed by computing the confidence limits

Lower limit	$LL = \overline{X} - m$
Upper limit	$UL = \overline{X} + m$

The confidence interval may be shown compactly as $\overline{X} \pm m$.

Suppose four samples are selected from the population and the four sample means $\overline{X}_1$, $\overline{X}_2$, $\overline{X}_3$, and $\overline{X}_4$ are computed (as well as the four sample standard deviations). These means are shown in Figure 14A.1(a). Applying the foregoing procedure, 0.95 confidence limits LL and UL are computed based on each sample and four confidence intervals are constructed, as shown in Figure 14A.1(b). The dotted vertical lines in the chart show the confidence interval related to each of the four sample means. The dot in the center of each confidence interval identifies the location of the sample mean. The solid vertical line shows the position of the population mean μ.

Notice that the widths of the confidence intervals vary, reflecting the different standard deviations s computed from the samples. The larger the value of s, the larger is the estimated standard error $s/\sqrt{n}$, the larger is the maximum expected error m, and the wider is the confidence interval.

Observe that the confidence intervals constructed based on $\bar{X}_1, \bar{X}_2, \bar{X}_3$, and the sample standard deviations include the population mean μ, whereas the confidence interval constructed based on $\bar{X}_4$ and the related sample standard deviation does not. Therefore, the first three confidence intervals provide useful estimates of μ, the fourth does not. If a large number of confidence intervals are constructed (based on a large number of samples) using the foregoing procedure with $z = 1.96$, then 95% of the confidence intervals may be expected to include μ and the other 5% will not.

Suppose a sample of 100 iron fittings is randomly selected from a population of 2,500 fittings. The sample fittings are found to have a mean weight of 8 ounces, with a standard deviation of 2.5 ounces. Construct a 0.95 confidence interval for the population mean weight.

First compute the estimated standard error of the mean as follows:

$$\frac{s}{\sqrt{n}} = \frac{2.5}{\sqrt{100}} = 0.25$$

Then, compute the fpc.

$$\text{fpc} = \sqrt{\frac{2500 - 100}{2500 - 1}} = 0.98$$

The z to be used for a 95% confidence interval may be determined from Table B.1 as 1.96. Then, compute the maximum error expected as follows:

$$m = 1.96 \times 0.25 \times 0.98 = 0.48$$

Finally, the confidence limits are

$$LL = 8 - 0.48 = 7.52 \text{ oz}$$
$$UL = 8 + 0.48 = 8.48 \text{ oz}$$

The 0.95 confidence interval for the mean is 7.52–8.48 oz.

Notice that the sampling ratio is 100/2500 = 0.04, which is less than 0.05. Therefore, the fpc may be ignored. Then, we compute m as follows:

$$m = 1.96 \times 0.25 = 0.49$$

The confidence limits become

$$LL = 8 - 0.49 = 7.51 \text{ oz}$$
$$UL = 8 + 0.49 = 8.49 \text{ oz}$$

Notice that the confidence interval computed ignoring the fpc (7.51–8.49 oz) is wider than when the fpc is included, so that including the fpc produces a better interval estimate. However, when the sampling ratio is less than 0.05, the improvement is negligible.

Note 14B

Constructing Confidence Intervals for a Population Proportion

This note presents a method for constructing a confidence interval for a population proportion where (1) the population is large, (2) the population proportion is not too close to 0 or 1, and (3) a sufficiently large sample is selected (at least 100, but a larger sample is desirable).

Note 13C discussed and described the sampling distribution of a proportion. This sampling distribution may be represented by the normal distribution under specified conditions, as discussed in **Note 13C**. The normal approximation is usually suitable in sampling problems encountered in practice.

The discussion of confidence intervals relating to a population mean in **Note 14A** is relevant to confidence intervals relating to a population proportion. In order to construct a confidence interval for a proportion based on sample data, we must first determine the sample size (n), the number of sample observations in the category of interest (X), and the sample proportion of observations in the category of interest (X/n). When using the sample proportion X/n as a point estimate of the population proportion p, we compute the maximum error expected (m) at a stated level of confidence as follows:

$$m = z \times \sqrt{\frac{\frac{X}{n} \times \left(1 - \frac{X}{n}\right)}{n}} \times \sqrt{\frac{N - n}{N - 1}}$$

The value of z depends on the confidence coefficient in a given problem (discussed in **Note 14A**). The factor $\sqrt{[X/n \times (1 - X/n)]/n}$ is the *estimated standard error of a proportion*. The third factor shown in the foregoing equation for m is the fpc (discussed in **Note 14A**).

The statistical theory underlying the equation for m (with the fpc ignored) is based on sampling from an infinite binomial population (or sampling with replacement from a finite binomial population). When sampling without replacement from a finite binomial population (the usual situation encountered in practice), the fpc must be included in the equation for m in order to make the adjustment required, because sampling is without replacement. If the sampling ratio (sample size/population size, or n/N) is less than 0.05, the fpc is so close to 1.00 that it may be ignored.

A confidence interval is constructed by computing the following limits:

$$\textit{Lower limit } LL = \frac{X}{n} - m$$

$$\textit{Upper limit } UL = \frac{X}{n} + m$$

The confidence interval may be shown as follows:

$$\frac{X}{n} \pm m$$

A random sample of 150 chemists selected from the full membership of 2,000 chemists in an association showed that 90 were well trained in statistical theory. Construct a 95% confidence interval for the population of the full membership who are well trained in statistical theory.

Based on the foregoing information we may write: $n = 150$, $X = 90$, $N = 2000$. We compute the relevant sample proportion as $X/n = 90/150 = 0.60$. The estimated standard error of a proportion is computed as follows:

$$\sqrt{\frac{\frac{X}{n} \times \left(1 - \frac{X}{n}\right)}{n}} = \sqrt{\frac{0.60 \times (1 - 0.60)}{150}} = 0.04$$

The fpc is:

$$\sqrt{\frac{2000 - 150}{2000 - 1}} = 0.96$$

Finally, since $z = 1.96$ for the 0.95 level of confidence, we compute the maximum expected error as follows:

$$m = 1.96 \times 0.04 \times 0.96 = 0.075$$

The confidence limits are as follows:

$$LL = 0.60 - 0.075 = 0.525$$
$$UL = 0.60 + 0.075 = 0.675$$

The 0.95 confidence interval for the population proportion is 0.525 – 0.675 (or 52.5% – 67.5%).

Note 14C

How to Determine Sample Size

When sampling, a decision must be made on the size of sample required. Sample size determination is often complicated since many factors must be considered, including the costs involved. We have developed sufficient statistical theory in **Notes 13B, 13C, 14A** and **14B** to permit a general consideration of how to estimate the size of sample to be selected in a given sampling situation.

Determination of sample size (n) depends on whether it is desired to estimate a population mean, a population proportion, or some other population characteristic. It must also be specified how accurate an estimate is required. That is, how much error is acceptable (since perfect estimates are not to be expected) and how much confidence is required that the maximum error associated with a point estimate does not exceed the specified amount.

If it is intended to estimate a population mean, then we know (**Note 14A**) that the maximum error to be expected at a stated level of confidence is

$$m = z \times \frac{\sigma}{\sqrt{n}}$$

Notice that we substituted the population standard deviation σ for its estimate (the sample standard deviation s) and omitted the fpc. Solving this equation for n, we obtain

$$n = \left(\frac{z \times \sigma}{m}\right)^2$$

How large a sample is required to estimate the mean score in a population of aptitude test scores based on sample data? It is specified that the estimate should be off (either way) by no more than 2 points at the 95% level of confidence.

Based on the information provided, we know that $m = 2$ points and $z = 1.96$. What should we use for σ, the population standard deviation? We do not need to know σ precisely for our purpose. A fair estimate may be available from a previous study. If not, we would have to collect a small pilot sample and compute s as an estimate of σ. Suppose σ is estimated as 18 points. Then, applying the foregoing equation,

$$n = \left(\frac{1.96 \times 18}{2}\right)^2 = 311.17 \text{ scores}$$

Therefore, a sample of 311 scores would be required. This result is not, of course, a precise measure of the sample size required, because σ is typically estimated. The general practice is to use a sensibly rounded figure for sample size (such as 300 scores or 325 scores).

If it is intended to estimate a population proportion, then we know (**Note 14B**) that the maximum error expected at a stated confidence level is

$$m = z \times \sqrt{\frac{p \times (1 - p)}{n}}$$

Notice that we substituted the population proportion p for its estimate (the sample proportion X/n) and omitted the fpc. Solving this equation for n, we obtain

$$n = p \times (1 - p) \times \left(\frac{z}{m}\right)^2$$

The use of the foregoing equation to determine sample size requires that an estimate of p is available. Such an estimate may be available from previous studies. If not, this equation may be modified to provide an estimate of the *maximum* sample size required. Notice that the foregoing equation includes the product $p \times (1 - p)$. It may be easily verified that this product has its maximum value when $p = \frac{1}{2}$. Then, $\frac{1}{2} \times (1 - \frac{1}{2}) = \frac{1}{4}$ is the maximum value of this product. Substituting $\frac{1}{4}$ for $p \times (1 - p)$ in the equation for n, we obtain

$$\text{maximum } n = \frac{1}{4} \times \left(\frac{z}{m}\right)^2$$

How large a sample of voters should be selected to estimate the proportion of the voters in a city who favor candidate Thomas? It is specified that the estimate should not be off by more than 0.05 (either way) at the 0.99 confidence level.

We know from the information given that $m = 0.05$ and $z = 2.58$ (for a 0.99 confidence level). If we have no idea as to the value to use for p, we may compute the maximum sample size needed as

$$\text{maximum } n = \frac{1}{4} \times \left(\frac{2.58}{0.05}\right)^2 = 665.64$$

Therefore, the largest sample needed is 666 voters. On the other hand, suppose it is generally felt that the proportion of voters who favor Thomas is about 0.30. Then, applying the earlier equation, we compute

$$n = 0.30 \times (1 - 0.30) \times \left(\frac{2.58}{0.05}\right)^2 = 559.14$$

Hence, if 0.30 is considered a fair estimate of p, a sample of 560 voters (instead of 666) is satisfactory.

If it is intended to construct a confidence interval for a population proportion using the procedure in **Note 14B**, sample size must be at least 100 (more is usually preferred) even if the sample size determined as just explained indicates a smaller sample.

15

Decision Making: A Statistical Approach

Administrators and other workers in business, industry, the professions, and government constantly face the task of decision making based on incomplete information. For example, a medical director may have to decide whether to replace an established surgical procedure with a newly developed one, or the manager of a large processing firm may have to decide whether to replace present equipment with something new and presumably better. The new procedure and the new equipment must be compared with the old. Although there may be adequate information relating to the good and bad points of the old surgical procedure and the old equipment, this is not generally true for the new procedure and equipment. Study and testing of the new procedure and equipment are required. However, this will provide only limited information. Decisions made on the basis of limited information carry the risk that unwise decisions will be made.

We will be concerned with decision-making problems where statistical methods can offer assistance in the decision process. This is the area of statistical inference which statisticians call *tests of hypotheses* or *tests of significance*. Let us consider some illustrations.

The superintendent of a large school system is considering the use of a new method to teach reading to first-year elementary-school pupils. This would involve considerable expense for new equipment. The superintendent knows that at the end of the first year, pupils obtain an average (mean) score of 115 on a standard reading achievement test. He decides to select a sample of pupils and try out the new teaching method for one year, using rented equipment. If the new method indicates that it will raise the mean score for the pupil population to at least 125, he will shift all schools to the new method and buy the new equipment. Otherwise, he will continue the use of the present method.

Let us consider the superintendent's problem. We know that a sample mean could vary considerably no matter what the value of the population mean, as shown in our discussion of Tables 12.2 and 12.4. Therefore, even if use of the new method would result in a mean score of 125 for the pupil population, *a particular sample mean could be quite different*. Suppose the sample mean turns out to be 123. Should the superintendent decide that the population mean may well be 125? Suppose the sample mean is 120. What decision should the superintendent make? If the sample mean turns out to be 118, is this different enough from 125 for the superintendent to conclude that the population mean score is less than 125? (*Even though such a sample mean could be obtained if the population mean actually is 125?*) This is a type of problem with which we are concerned.

Consider the problem facing the factory manager in a large firm who is considering the purchase of a new testing machine. The producer of this machine states that, due to the sensitivity of the testing mechanism, the machine could be expected to jam from time to time. However, he claims that it will jam no more than 15% of the time and that a touch of a lever will restore the machine to normal working condition. The manager is very interested in the new machine because it represents a great improvement over his present testing equipment. However, it is very expensive and he would not be interested in buying it if it jammed more often than 15% of the time.

The factory manager decides to ask the producer of the machine to permit him to test it. He plans to make 500 trials of the machine and the proportion of times the machine jams during this sample of trials will be used to evaluate its performance. Clearly, the problem facing the factory manager is not unlike the school superintendent's problem. That is, even if the proportion of times the machine will jam over a long period of use (the population of trials) is 0.15 (or 15%), *the sample proportion could be quite different*.

Consider a final illustration. Are men and women equally productive in wiring radio circuits? In order to evaluate productivity, a random sample is selected from the population of male wirers and a random sample is selected from the population of female wirers. Then, the average (mean) number of circuits wired per worker is computed for each sample. The *difference between the two sample means* will provide the basis for evaluating the *difference between the two population means* (average productivity of all male and all female wirers). However, even if the average number of circuits wired is the same for the population of male wirers and the population of female wirers, *the difference between two sample means could be substantial*. Here, too, we face the problem of making a decision based on limited information, with the risk that an incorrect decision could be made.

15.1 Statistical Decision Making

Generally speaking, statistical decision making may be looked upon as made up of three phases, as follows:

Phase A *Formulating the problem.* This involves identifying a measurable characteristic pertinent to the problem, formulating hypotheses related to the alternative decisions which could be made, and specifying how much risk may be taken that a wrong decision will be made.

Phase B *Formulating the decision rule.* This takes into account the various aspects of the decision-making problem formulated in *Phase A* and specifies precisely how the decision will be made.

Phase C *Evaluation of the sample data.* Computations based on the sample are referred to the decision rule (*Phase B*) and the decision is made.

Let us consider each phase more carefully.

15.2 Formulating the Problem and Decision Risks

Notice that in each of the illustrative problems at the beginning of this chapter, the decision to be made was reformulated in terms of a *measurable characteristic.* The school-system superintendent recast his problem relating to adoption of the new teaching method in terms of the *mean pupil score* on an achievement test (a measurable characteristic). The factory manager recast his problem relating to purchase of a new machine in terms of the *proportion of times the machine may be expected to jam* (a measurable characteristic). The problem relating to comparative productivity of men and women in wiring radio circuits was recast in terms of the *difference in the mean number of circuits wired per worker* between men and women (a measurable characteristic).

Statistical decision making requires a precise statement of the problem and determination of the alternative decisions that could be made. In the foregoing illustration of the factory manager, the problem "to purchase or not to purchase the new machine" cannot be handled statistically. However, the more precise formulation relating to the proportion of times the machine may be expected to jam can be handled statistically. We think of all possible trials of the machine during its expected lifetime as a population of trials. Then, the proportion of these trials resulting in a jam is the *parameter* (a characteristic of a population) to be used in the decision-making procedure.

A key factor in statistical decision making is the formulation of *appropriate hypotheses.* We will consider only the simplest situation where two hypotheses are formulated—the *hypothesis to be tested* and an *alternative hypothesis.* The hypothesis to be tested evolves from the formulation of the decision-making problem. In the case of the factory manager, the decision to be made was reformulated in terms of the proportion of times the machine may be expected to jam. Is this proportion 0.15 as claimed by the producer of the machine or is it greater than 0.15? Consider the possible hypotheses:

(a) Proportion is greater than 0.15
(b) Proportion $= 0.15$

Statistical theory requires that the hypothesis to be tested is *specific*. The first of the foregoing hypotheses is not specific since any proportion greater than 0.15 satisfies this hypothesis. The second hypothesis is a specific hypothesis since only one proportion (0.15) satisfies it. Then, this is an appropriate hypothesis for statistical testing. The factory manager will hypothesize that the claim of the producer of the testing machine is true (*the proportion of trials which result in a jam is 0.15*). This does not mean that he necessarily believes it is true. It means only that this is the hypothesis he will test.

The factory manager obtains a hypothesis appropriate for testing (a specific hypothesis) if he hypothesizes that there is *no difference* between the producer's claim and the true proportion of trials which will result in a jam of the machine. Such a hypothesis statisticians call a *null hypothesis*. Generally speaking, a *null hypothesis* is a hypothesis of no difference between the true value of a population parameter and a claimed value.

Statistical testing of the null hypothesis may lead to either the *rejection* of this hypothesis or its *acceptance*. If rejected, an *alternative hypothesis* will be accepted. An appropriate alternative hypothesis must be carefully formulated to reflect the realities of the decision-making problem. Recall that the factory manager is interested in evaluating the claim that the proportion of trials which result in a jam is 0.15 and, if this proportion is *greater than* 0.15 (making the null hypothesis false), he will reject the null hypothesis and decide not to purchase the machine. Clearly, he is not concerned with the possibility that the proportion is *less than* 0.15, since he has no doubt that the producer of the machine is making the best claim he can for his product. Hence, the appropriate alternative hypothesis in this decision-making problem is "Proportion is greater than 0.15."

Then, the statistical test will proceed on the basis that if the hypothesis that the machine will jam 15% of the time is rejected, the factory manager can only conclude that it will jam more often.

The features of a statistical test which we have been discussing, as well as some other features, are presented in Figure 15.1. Here we show that one of two decision routes may be taken in a decision-making problem.

1. If the null hypothesis is rejected, accept the alternative hypothesis.
2. If the null hypothesis is not rejected, accept it.

Clearly, whichever choice is made, there is the risk that an error (an incorrect decision) will be made. Statisticians identify two types of errors that may be committed. If the null hypothesis is true, but the test leads to rejection of this hypothesis, we make a *type I error*. Such an error is possible if the test leads to adoption of the first of the foregoing decision routes. On the other hand, if the null hypothesis is false, but the test does not lead to rejection and it is accepted, a *type II error* is committed. Such an error is possible if the test leads to adoption of the second decision route. The error possibilities are shown in the circles in Figure 15.1.

An important contribution of statistics to decision making is the fact that *the risks that an error will be made may be predetermined*. For example, say the factory manager decides to conduct a statistical test. Then, as just noted, he exposes himself to the

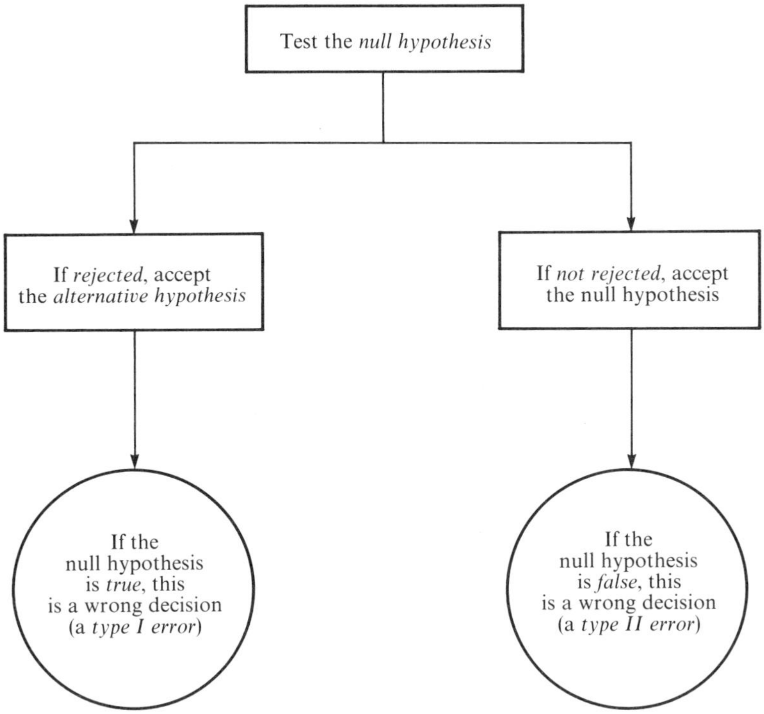

Figure 15.1

Some Features in a Statistical Test of a Hypothesis

risks of making a type I or a type II error. The factory manager would evaluate the seriousness of each type of error and then decide (perhaps with the statistician's assistance) how much risk he wishes to take that each type of error will occur. He recognizes that he would commit an error (type I) if the producer's claim is true, but the test leads him to reject the claim "accept the alternative hypothesis that the proportion is greater than 0.15," and decide not to purchase the machine. He may not consider this error too serious because his present equipment is performing well enough and may wish to take a risk as high as 5 in 100 (a probability of 0.05) that such an error could occur. On the other hand, he may feel that the new testing machine would reduce production costs considerably and it is important to avoid such an error from occurring. Then, he would want to take only a small risk, perhaps 0.02 (2 chances in 100), 0.01 (1 chance in 100), or even 0.005 (5 chances in 1,000) that such an error would occur.

If the producer's claim is actually false, the factory manager would commit an error (type II) if the test does not lead him to reject this claim and he decides to purchase the machine. Since the machine is expensive, he would consider such an error serious indeed and would want to take a very small risk (0.005 or 0.01) of making it. The statistician would then incorporate the risks specified by the factory manager when formulating the decision rule (phase B).

Very often, for technical reasons, only the risk of a type I error is predetermined. This may be considered the more important risk to be protected against. Statisticians call the probability of committing a type I error the *level of significance* of the test.

15.3 Decision Rule and Sample Data Evaluation

It will be recalled that a parameter is a characteristic of a population and which parameter is to be used in a test is determined when a decision-making problem is formulated. In the tests we have been discussing, the parameter is estimated by computing a *statistic*, which is a corresponding measure based on the sample data. For example, if the *population mean* is the parameter, the *sample mean* is the corresponding statistic. If a *population proportion* is the parameter, the corresponding *sample proportion* is the statistic.

Formulation of a *decision rule* is based on the *sampling distribution of the statistic used in a test*. For example, if the statistic is the *sample mean*, then the decision rule is based on the *sampling distribution of the mean* (as presented in Tables 12.2 and 12.4). If the statistic is a *sample proportion*, then the decision rule is based on the *sampling distribution of a proportion* (discussed in connection with Table 13.1). The *decision rule* specifies when to reject or accept the null hypothesis.

Let us further consider the factory manager's problem. Recall that he plans to make 500 trials of the testing machine (a sample of trials) and the proportion of times the machine jams in the sample is the statistic to be used as a basis for deciding whether or not to purchase the machine. The concern of the manager is that the proportion may be greater than 0.15 (as expressed in the alternative hypothesis). How much greater than 0.15 must the sample proportion be for the manager to decide to reject the producer's claim? This depends on *how much risk* he is willing to take that such a decision may be wrong.

The closer the sample proportion to the producer's claim of 0.15, the less likely it is that the claim is false and the greater the risk that rejection of this claim is an error. On the other hand, the larger the sample proportion (and so the further away it is from the producer's claim), the more likely it is that the claim is false and the smaller the risk that rejection of this claim is an error.

It has been noted that the sampling distribution of a proportion may be approximated by a normal distribution. Suppose the factory manager decides to conduct the test at the 0.05 level of significance. As previously noted, this means that he is willing to take a risk of 5 in 100 that he will make the error of rejecting the producer's claim if it is true (type I error). Applying this significance level, and using the normal distribution to approximate the sampling distribution of a proportion, the statistician can determine the sample proportion beyond which the manager should decide to reject the null hypothesis. This proportion, determined by the statistician for the factory manager's problem, is 0.176. We call this proportion the *critical value*. Then, the statistician can construct the following *decision rule*:

Decision rule Compute the proportion of times the machine jams in the sample of 500 trials. If this proportion is greater than 0.176, reject the producer's claim, accept the alternative hypothesis that the proportion is greater than 0.15, and decide not to purchase the machine. Otherwise, accept the producer's claim that the proportion is 0.15 and purchase the machine.

The foregoing decision rule was constructed so that the probability is 0.05 (5 chances in 100) that the producer's claim will be rejected if it is true (a type I error). Even if the population proportion is actually 0.15, a sample proportion could have a wide range of values (as indicated by the sampling distribution of a proportion). The critical value 0.176 was determined so that 0.05 (or 5%) of these possible sample proportions are larger than the critical value. In other words, the range of possible sample proportions is divided into two parts: the sample proportions greater than the critical value which make up a *region of rejection*, and the other sample proportions which make up a *region of acceptance*.

The decision rule specifies that if the sample proportion falls in the region of rejection (also called the *critical region*), the hypothesis being tested will be rejected and the alternative hypothesis will be accepted. If the sample proportion falls in the region of acceptance, the null hypothesis will be accepted.

Suppose the factory manager finds that 92 out of the sample of 500 trials results in a jam of the machine. Then, the sample proportion is $92/500 = 0.184$. Referring to the decision rule, we note that 0.184 is greater than 0.176 and so falls in the critical region. Consequently, the producer's claim is rejected and the decision is made not to purchase the machine.

What does it mean if the null hypothesis is rejected? It does *not* mean that the null hypothesis has been *proven false*. It means that it is *very improbable* that the sample proportion obtained would have been obtained if the population proportion were actually 0.15. However, this is *not impossible* and the factory manager is taking a risk of 0.05 (5 chances in 100) that he has made an incorrect decision (type I error).

A statistical test cannot prove whether a null hypothesis is true or false. Based on the sample data, the hypothesis tested is rejected or not rejected. If rejected, this indicates that it is unlikely that the null hypothesis is true and so the alternative hypothesis is accepted. If not rejected, this indicates that the hypothesis tested has not been shown to be improbable and so it is accepted.

In subsequent chapters, we will examine various types of decision-making problems and observe how statistical decision making is applied. This will help the reader to better understand this important statistical procedure.

Review

Concepts and Terminology

Acceptance of a hypothesis
Alternative hypothesis
Chance outcome
Critical value
Hypothesis to be tested
Level of significance of a test
Null hypothesis
Parameter

Concepts and Terminology (*cont.*)

Decision errors
Decision making based on incomplete information
Decision risks
Decision rule
False hypothesis
Formulating a decision problem
Hypotheses
Hypothesis of no difference
Predetermining errors
Rejection of a hypothesis
Specific hypothesis
Statistic
Tests of hypotheses
Tests of significance
True hypothesis
Type I error
Type II error

Problems

1. Define and illustrate: parameter, statistic.

2. Indicate the parameter and the corresponding sample statistic.
- (a) A purchaser of light bulbs is considering the purchase of Company A bulbs and will do so if the average life of these bulbs is not less than 1000 hours.
- (b) A cosmetic door-to-door salesman claims that he usually obtains orders from a high proportion of calls.
- (c) A feed grain manufacturer claims that a new feed concentrate will be more effective than a standard feed in fattening a pig (on the average).
- (d) It is claimed that a certain college course attracts more female students than male students.
- (e) It is claimed that college freshmen drop out of college at a higher rate than college sophomores.
- (f) It is claimed that a certain plant variety grows to a mean height of 4.2 centimeters when grown in the western part of a state.
- (g) A law firm claims that 60% of the cases it handled in the last two years were settled out of court.

3. Select the hypothesis appropriate for a test of hypothesis and state why it is appropriate.
- (a) Part (a) of Problem 2:
  1. Mean bulb life = 1,000 hours
  2. Mean bulb life is less than 1000 hours
- (b) Part (b) of Problem 2:
  1. Proportion of calls which result in an order = 0.80
  2. Proportion of calls which result in an order is less than 0.80
- (c) Part (c) of Problem 2:
  1. Mean weight gain per pig for the new feed = mean weight gain for the standard feed
  2. Mean weight gain per pig for the new feed is greater than for the standard feed
- (d) Part (d) of Problem 2:
  1. Proportion of students who enroll for the course who are females = 0.50
  2. Proportion of students who enroll for the course who are females is greater than 0.50
- (e) Part (e) of Problem 2:
  1. Proportion of freshmen who drop out of college = proportion of sophomores who drop out of college

2. Proportion of freshmen who drop out of college is greater than the proportion of sophomores who drop out

(f) Part (f) of Problem 2:
1. Mean height does not equal 4.2 centimeters
2. Mean height = 4.2 centimeters

(g) Part (g) of Problem 2:
1. Proportion of cases handled out of court is not 0.60
2. Proportion of cases handled out of court = 0.60

4. "Even if the mean weight for a population of objects is 250 pounds, it is possible for the mean weight computed for a sample of these objects to be different." Explain why this is true.

5. "If the mean for a population of prices is \$180, it is impossible for the mean of a sample of these prices to equal \$50 or less." True or false? Defend your answer.

6. The mayor of a large city claimed that 80% of the voters favored a new tax he was proposing. A sample of voters showed that 20% favored the new tax proposal. Does this indicate that it is impossible that the mayor's claim is true? Why?

7. What is meant by a null hypothesis?

8. Formulate the hypothesis to be tested and an appropriate alternative hypothesis.
(a) A drug manufacturer claims that a new product will cure 70% of the people who have a certain disease. A hospital thinks the cure rate is less than 70% and wishes to conduct a statistical test to evaluate the claim.
(b) The seller of a large supply of metal rings claims that no more than 20% of the rings are out of shape. The buyer thinks that more than the claimed percentage is out of shape and wishes to conduct a statistical test based on a sample to evaluate the claim.
(c) The average diameter of a large supply of apples is claimed to be 2 inches. A buyer believes that it is more than 2 inches and plans to conduct a statistical test to evaluate the claim.

9. Distinguish between a type I and a type II error.

10. Refer to part (a) of Problem 8. Indicate how a type I and a type II error could occur.

11. What is meant by the level of significance of a statistical test of hypothesis?

12. What is meant by a decision rule?

13. What is the critical value of a statistic and how is it used in conducting a test of hypothesis?

14. When a null hypothesis is rejected, it may be that this hypothesis is actually true. Why?

15. A seller offered a carload lot of peaches for sale at a discount price. He claimed that 5% of the peaches were infected by brown rot. A potential buyer is interested in buying the full supply, but wants first to feel convinced that no more than 5% of the supply of peaches is infected by brown rot. A sample of 100 peaches included 35 that are affected. Does the sample evidence prove that 5% is a dishonest estimate? Why?

16. In Problem 15, the potential buyer decides to conduct a statistical test of the supplier's claim.
(a) Formulate the appropriate hypothesis to be tested and the appropriate alternative hypothesis.
(b) If the supplier's claim is true, but the test leads the potential buyer to reject it, what type of error is this?
(c) Suppose the potential buyer considers the discount price very low and would very much like to make the purchase. Would a 0.05 or a 0.01 risk of making the error of rejecting the supplier's claim, if it is true, be more suitable? Why?

(d) If the supplier's claim is false, but the test leads the potential buyer to purchase the supply of peaches, what type of error is this?

(e) Suppose the potential buyer considers the discount price not too attractive and is only mildly interested in making the purchase, even if the supplier's claim is true. Would a 0.05 or a 0.005 risk of rejecting the supplier's claim, if it is true, be more suitable? Why?

17. A senior partner in a law firm claimed that an average of 30 days per case was spent on cases handled by the firm during the past five years. He felt that because of this the number of lawyers hired by the firm should be increased. Another senior partner felt that the average time spent on a case was less than 30 days. Suppose it was decided to conduct a statistical test of this claim.

(a) Formulate the appropriate hypothesis to be tested and the logical alternative hypothesis.
(b) How could a type I error be made?
(c) How could a type II error be made?

18. In Problem 17:

(a) Suppose the firm is very much against hiring more lawyers unless the claim is true. Would a probability of 0.05 or 0.005 be more suitable as protection against making a type II error?

(b) Suppose the firm were thinking about increasing the lawyer staff, whether or not the claim is true. Would a 0.05 or 0.01 probability of making a type I error be more suitable?

19. A test of hypothesis is to be conducted relating to the proportion of apples in a large supply which are infected with apple scab. The decision rule states: If the proportion of apples in a sample which have apple scab is greater than 0.10, reject the claim that only 8% of the apples are infected, accept the alternative hypothesis that the population proportion is greater than 8%, and decide not to purchase the supply of apples. Otherwise, accept the claim and purchase the supply of apples.

(a) What is the critical value in this decision rule?
(b) Identify the critical region.
(c) Identify the region of acceptance.
(d) Suppose the probability of a sample proportion falling in the region of rejection is 0.01. What is the level of significance of the test?

20. Consider the following decision rule: If the sample mean is less than $500, reject the hypothesis that the population mean is $800 and accept the alternative hypothesis that the population mean is less than $800. Otherwise, accept the hypothesis being tested (that the population mean is $800).

(a) Identify the critical value.
(b) Identify the region of acceptance.
(c) Identify the region of rejection.
(d) Suppose the level of significance of the test is 0.02. What is the probability that a sample mean will fall in the region of rejection if the null hypothesis is true?
(e) Suppose the level of significance of the test is 0.05. What is the probability of committing a type I error?

16

Decision Making: Evaluating a Claim

16.1 Evaluating a Proportion

A coding clerk tells her supervisor that she is thinking of accepting another job, but will stay on if she can get a raise in salary. The supervisor tells her that he will consider the matter and let her know in a week. He would not like to lose this employee; however, he wants to be sure that she does better work than the others before he decides to raise her salary. Otherwise, it would create a morale problem.

The supervisor decides to select a sample of 600 codes from the employee's work during the previous month in order to test the accuracy of her work compared to the others. He knows that 5% of the codes prepared by the group of coding clerks are found to be incorrect. He decides to take only a small risk (0.01 or 1 chance in 100) that he will be mistaken in concluding that this clerk makes less than 5% errors.

We will apply the testing procedure discussed in the previous chapter. The following hypotheses are appropriate to this problem:

1. *Hypothesis to be tested (the null hypothesis)*. Proportion of codes prepared by the clerk that are incorrect = 0.05 (the same as for the coding clerks considered as a group)
2. *Alternative hypothesis*. Proportion is less than 0.05

Notice that the alternative hypothesis states that the proportion is less than 0.05, because the supervisor is interested in evaluating the possibility that this clerk makes *fewer errors* in coding. The possibility that she makes more errors (proportion is greater than 0.05) is not pertinent to this problem. Since the alternative hypothesis is concerned only with proportions *less than* stated in the null hypothesis, statisticians call this a *one-sided hypothesis* and refer to such a statistical test as a *one-tail test* or more specifically as a *lower-tail test*. Recall the factory manager's problem in the previous chapter. Here, we had the alternative hypothesis "The proportion is greater than 0.15." This is also a one-sided alternative hypothesis and the test is a one-tail test, or more specifically, an *upper-tail test*, because it is concerned only with proportions *greater than* 0.15 (the null hypothesis).

The level of significance for this test is 0.01, as specified by the supervisor. Based on the level of significance and the hypotheses formulated, we derive the following decision rule:

Decision rule If the proportion of errors in the sample of 600 codes is less than 0.029, reject the null hypothesis, accept the alternative hypothesis, and raise the coding clerk's salary. Otherwise, accept the null hypothesis and do not raise the clerk's salary.

The statistician determined that even if the clerk's error rate is 0.05, the probability is 0.01 (the level of significance of this test), that the proportion of errors in a particular sample could be *below* 0.029. Hence, 0.029 was selected as the critical value as noted in the decision rule. Suppose the sample of 600 codes includes 16 incorrect ones. Then, the sample proportion of errors is $16/600 = 0.027$, which is less than the critical value. The supervisor, in accordance with the decision rule, decides to raise the clerk's salary. (*Note*: This does not *prove* that the null hypothesis is false. Since the sample proportion 0.027 falls in the region of rejection, it indicates that it is *very unlikely* that this hypothesis is true. Therefore, the null hypothesis is rejected and the alternative hypothesis is accepted.)

Hank Smith, a candidate for governor, claimed that he could count on 60% of the votes in the forthcoming election. One opponent stated that if he thought Smith could claim less than 60% of the votes, he would throw all his resources into a campaign against him. Otherwise, he felt this would not be a wise move. Another opponent felt that Smith could count on more than 60% of the votes and was thinking of withdrawing from the race. A voter-research committee decided to select a random sample of 100 voters to test candidate Smith's claim and to take a risk of 2 chances in 100 (0.02) that they would erroneously decide that the claim was false. The appropriate hypotheses are as follows:

1. *Hypothesis to be tested.* Proportion of voters who favor candidate Smith = 0.60
2. *Alternative hypothesis.* Proportion is not 0.60 (it is more than 0.60 or less than 0.60)

Notice that the alternative hypothesis states that the proportion is not 0.60, indicating that it could be on either side of the null hypothesis (more than 0.60 or less than 0.60). In other words, this test seeks to determine whether the proportion of voters who favor candidate Smith is, or is not, 0.60. Statisticians call this a *two-sided hypothesis* and refer to such a test as a *two-tail test*. The level of significance of this test is 0.02, as determined by the voter-research committee.

Based on the level of significance and the hypotheses formulated, the statistician specifies the following decision rule:

Decision rule If the proportion of voters in the sample who favor Smith is either less than 0.486 or more than 0.714, reject the null hypothesis, accept the alternative hypothesis, and conclude that the proportion of voters who favor Smith is not 0.60. Otherwise, conclude that the proportion is 0.60.

The statistician determined that even if the proportion of voters favoring Smith is 0.60, the probability is 0.02 (the level of significance) that in a particular sample this proportion could be either less than 0.486 or more than 0.714. Therefore, these were selected as the critical values in the decision rule.

Notice that in a two-tail test there are *two critical values*, a *lower-tail critical value* (0.486) and an *upper-tail critical value* (0.714), because it is necessary to evaluate differences from the null hypothesis in *both directions.* More specifically, if the sample proportion who favor Smith is below 0.486 or above 0.714 (either way), it is different enough from 0.60 to warrant rejection of the null hypothesis.

It should be clear that in a one-tail test interest is in differences from the null hypothesis in only one direction; whereas, in a two-tail test interest is in differences in either direction. (**Note 16A** at the end of the chapter shows how to construct a test of a hypothesis relating to a population proportion.)

Suppose 69 out of the 100 voter sample favor candidate Smith. Then, the sample proportion favoring Smith is $69/100 = 0.69$, which is larger than the null hypothesis (0.60) but not larger than the upper-tail critical value (0.714). The voter research committee, in accordance with the decision rule, decides to accept candidate Smith's claim.

16.2 Evaluating a Mean

A machine was set to cut aluminum wire with a mean length of 35 inches. Before putting the machine into production, it was decided to examine the first 50 cuts to determine whether or not the mean length of cuts was as desired. It was found that the mean length of the sample of cuts was 36.42 inches, with a standard deviation of 6.30 inches. Would you conclude, at the 0.05 level of significance, that the mean length per cut was as desired? The hypotheses for this test are as follows:

1. *Hypothesis to be tested.* Mean length of cuts = 35 in
2. *Alternative hypothesis.* Mean length is not 35 in

Decision rule If the mean length of cut for the sample is either less than 33.26 inches or more than 36.74 inches, reject the null hypothesis, accept the alternative hypothesis, and conclude that the mean length is not 35 inches. Otherwise, conclude that it is 35 inches.

In other words, the statistician determined that even if the mean length per cut is 35 inches, the probability is 0.05 (the level of significance) that in a particular sample it could be either less than 33.26 inches or more than 36.74 inches. Therefore, these two values were selected as the critical values in the decision rule.

Since the sample mean length was found to be 36.42 inches, just below the upper-tail critical value, it was concluded that the mean length of cut is 35 inches. The machine is put into production. This decision follows from the decision rule.

A social psychologist was making an in-depth study of social attitudes in a large community. He felt from his investigations that, on the average, the adults in the community were below the national level in racial tolerance. He selected a sample of 22 adults from the community to test whether they would score sufficiently below the nation-wide mean of 125 on a racial tolerance test, to warrant the conclusion that his feelings about the adults in the community were correct. He found that the sample obtained a mean score of 121.3 on the test, with a standard deviation of 12.7 points. Using the 0.05 level of significance, should he conclude that his feelings are correct? The following hypotheses were formulated:

1. *Hypothesis to be tested.* The mean test score for the adults in the community = 125
2. *Alternative hypothesis.* The mean is below 125

Decision rule If the sample mean test score is less than 120.3, reject the null hypothesis, accept the alternative hypothesis, and conclude that the population of adults, on the average, is below the national level in racial tolerance. Otherwise, conclude that the adults in the community on the average are not below the national level.

Noting that the sample mean score of 121.3 is higher than the critical score of 120.3, the social psychologist concluded that his feelings about the adults in the community were not correct. This decision was made in accordance with the decision rule.

(**Note 16B** at the end of the chapter shows how to construct a test of hypothesis relating to a population mean.)

16.3 Evaluating Variation

A machine was set to produce fittings at a specified rate of output per minute, with a variation in output per minute no greater than a standard deviation of 3 units. If the output variation exceeded the 3 units standard deviation, the machine would have to be adjusted. A sample of 15 minutes of output showed a standard deviation of 3.5 units. Using the 0.05 level of significance, should it be concluded that the machine requires adjustment? The appropriate hypotheses for this test are as follows:

1. *Hypothesis to be tested.* The standard deviation for output per minute = 3 units
2. *Alternative hypothesis.* The standard deviation is more than 3 units

Decision rule If the sample standard deviation exceeds 3.9 units, reject the null hypothesis, accept the alternative hypothesis, and conclude that the standard deviation of output per minute is more than 3 units and that the machine needs adjustment. Otherwise, conclude that the machine does not need adjustment.

Since the sample standard deviation (3.5 units) is below the critical value, it is concluded that the machine does not need adjustment. This is in accordance with the decision rule.

(The decision rule in the foregoing illustration is based on a new concept, the *chi-square statistic*, which is related to a theoretical distribution called the *chi-square distribution*. **Note 16C** at the end of this chapter discusses this distribution. **Note 16D** shows how to construct a test of hypothesis relating to the standard deviation of a population.)

Review

Concepts and Terminology

Lower-tail critical value
Lower-tail test
One-sided hypothesis
One-tail test
Two-sided hypothesis
Two-tail test
Upper-tail critical value
Upper-tail test

Problems

1. A school superintendent claims that the mean reading level of third graders in his school system is 120 on a certain reading ability test. A specialist wants to test this by administering a new form of the same test to a sample of third graders. The specialist has no preconceived idea that the claim is not true. He plans to conduct the test with a probability of 0.005 that he will make an error if he concludes that the claim is not true.
(a) Formulate the hypothesis to be tested and the appropriate alternative hypothesis.
(b) What is the level of significance of this test? Explain what it means.
(c) What is the probability of a type I error for this test?
(d) How many critical values are required for this test? Why?

2. In Problem 1, formulate the hypothesis to be tested and the appropriate alternative hypothesis if the educational specialist believes that the mean reading level score is less than 120.

3. The decision rule in a test of hypothesis is as follows: If the mean amount of deposit withdrawals from savings and loan associations for a random sample of 150 depositors is less than $175.25 or more than $184.75, reject the hypothesis that the mean withdrawal for the population of savings and loan accounts is $180 and accept the alternative hypothesis that the mean is not $180. Otherwise, conclude that the mean amount withdrawn for the population of accounts is $180.
(a) What is the null hypothesis for this test?
(b) What is the critical value (or values) for this test?
(c) If a sample of 150 depositors shows a mean withdrawal of $185, what should be concluded concerning the mean amount of withdrawals in the population? Why?

4. In Problem 3, identify the regions of acceptance and rejection.

5. A biologist claims that the hydras in a large tank have a mean length of 5 millimeters. A hydra is a tiny sea animal found in fresh water. The senior biologist believes that the mean length is larger and plans to select a sample of the hydras in the tank and conduct a statistical test. He decides to design the test so that there is a probability of 0.05 that he will be in error if he rejects the claim of the biologist and the claim is true.
(a) Formulate the hypotheses for this test.
(b) What is the level of significance of this test and what does it mean?
(c) How many critical values are required for this test? Why?

6. Partner A in a law firm claims that a mean of 20 days per case must be spent in court and thus fees per case should be increased. Partner B thinks that the mean number of days is less than 20. Suppose Partner B has a statistical test conducted using the 0.01 level of significance.
(a) Formulate the hypotheses for this test.
(b) How many critical values are required for this test? Why?

7. In Problem 6—
(a) Which sampling distribution is the basis for the decision rule?
(b) Identify the critical region.
(c) Identify the region of acceptance.
(d) What is the probability that the sample of cases selected for the test will have a mean number of court days which fall in the critical region if the null hypothesis is true? Why?

8. In Problem 6, suppose Partner B is interested in testing whether or not the claim of Partner A is a valid one, with no strong feelings that it is either too high or too low.
(a) Formulate the appropriate hypotheses.
(b) How many critical values are required for this test? Why?
(c) How many rejection regions should be included in the decision rule? Identify them.
(d) Identify the region of acceptance.
(e) What is the probability that the sample of cases selected for the test will have a mean number of court days which will lead to rejection of Partner A's claim if the null hypothesis is true? Why?

9. A seller of small steel balls claims that the mean weight of a ball is 6.8 ounces. A potential buyer decides to select a sample of 400 balls and conduct a statistical test with the following null hypothesis: Mean weight = 6.8 ounces. The following is the alternative hypothesis: Mean weight is more than 6.8 ounces. The level of significance is 0.02. The decision rule is: If the mean weight of the sample is more than 7.0 ounces, reject the null hypothesis and conclude that mean weight per ball is more than 6.8 ounces. Otherwise, conclude that the mean weight is 6.8 ounces.
(a) If the sample shows a mean weight of 7.2 ounces, what should be concluded based on the decision rule?
(b) What is the sampling distribution on which the decision rule is based?

10. An electric shaver firm claims that 25% of the men who use electric shavers prefer their product. A consumer study organization doubts that this claim is true (it is thought to be too high) and decides to select a random sample to test the claim. The test is to be conducted with a 5% probability of being wrong, if the claim is not accepted.
(a) Formulate the hypothesis to be tested and the appropriate alternative hypothesis.
(b) What is the level of significance of this test and what does this mean?
(c) How many critical values are required for this test? Why?

11. The decision rule in a test of hypothesis is: If the proportion of defective fittings in a random sample of 500 selected from a large supply offered for sale is more than 0.126, reject the hypothesis that the proportion = 0.10, accept the alternative hypothesis that the proportion is greater than 0.10, and decide not to purchase the entire supply of fittings. Otherwise, buy the supply of fittings.
   (a) What is the null hypothesis in this test?
   (b) What is the critical value for this test?
   (c) If 55 fittings in the sample are found to be defective, what decision should be made? Why?

12. A medical researcher claims that 75% of the adults in the country have been given at least one antibiotic shot to help combat a disease. A sample of adults is randomly selected to test this claim, using the 0.05 level of significance. The test is to be conducted to verify whether or not this claim is true.
   (a) Formulate the appropriate hypotheses for the test.
   (b) How many critical values are required for this test? Why?

13. In Problem 12, assume the test is conducted because it is felt that 75% is too high. The significance level of the test is changed to 0.02.
   (a) Formulate the hypotheses for the test.
   (b) How many critical regions are required for this test? Why?
   (c) What is the probability that the proportion computed from the sample data will fall in the critical region if the claim is actually true?
   (d) If the level of significance remained 0.05 as originally stated, what is the probability you would have determined in your answer to (c)?
   (e) Is the rejection region larger if the level of significance is 0.02 or 0.05? Why?

14. A store manager claims that 40% of his customers are teenagers. A random sample of customers is to be selected to test this claim because it is felt that it is too high. The 0.02 level of significance is specified for the test.
   (a) Formulate the appropriate hypotheses.
   (b) Is the alternative hypothesis one-sided or two-sided? If one-sided, is it a lower-tail test or an upper-tail test?

15. In Problem 14:
   (a) Should the critical value lie below the parameter value specified in the null hypothesis or above it? Why?
   (b) Which sampling distribution is to be used to construct the decision rule?
   (c) What proportion of possible sample proportions represented by the sampling distribution will lie below the critical value? Why?
   (d) Is the part of the sampling distribution which lies below the critical value the acceptance region or rejection region?

16. The mayor of a large city announced that he would support a new housing bill because 80% of the city's residents favored it. The legislature did not think the percentage was that high. A random sample of 600 voters was selected and a statistical test conducted, using the 0.05 significance level. The null hypothesis is: Proportion in favor = 0.80. The alternative hypothesis is: Proportion in favor is less than 0.80. The decision rule is: If the sample proportion in favor is less than 0.54, reject the null hypothesis and conclude that fewer than 80% are in favor. Otherwise, conclude that 80% are in favor.
   (a) What is the critical value in this test?
   (b) Identify the region of rejection.
   (c) What is the sampling distribution on which the decision rule is constructed?

(d) What is the probability that a sample proportion will fall in the rejection region if the null hypothesis is true?
(e) Identify the region of acceptance.
(f) It was found that 250 residents in the sample favor the housing bill. Apply the decision rule and come to a conclusion.

17. A department store manager claims that the time required to handle customer complaints has a standard deviation of 10 minutes. The president of the firm believes this is an overstatement and asks his accountant to conduct a statistical test of this claim.
(a) Formulate the hypothesis to be tested and the appropriate alternative hypothesis.
(b) How many critical values are required for this test? Why?

18. A librarian claims that the number of books borrowed per day is fairly stable and has a standard deviation of 20 books. The library director believes that the standard deviation is higher and decides to conduct a statistical test based on books borrowed for a random sample of days. He decides to use the 0.05 level of significance for the test.
(a) Formulate the hypotheses for the test.
(b) How many critical values are required for this test? Why?
(c) What is the probability that the sample data will lead to rejection of the null hypothesis if the librarian's claim is true?

19. In Problem 18, suppose the level of significance is changed to 0.01 and the library director conducts a two-tail test to evaluate whether or not the claim is true.
(a) Formulate the appropriate hypotheses.
(b) How many critical values are required for the decision rule? Why?

20. The alternative hypothesis determines the location of the region of rejection in the appropriate sampling distribution (lower-tail, upper-tail, both tails). In Problem 9, where is the critical region located based on the following:
(a) The alternative hypothesis as stated
(b) *Alternative hypothesis.* Mean weight is less than 6.8 ounces
(c) *Alternative hypothesis.* Mean weight is not equal to 6.8 ounces

21. In Problem 16, where is the rejection region located based on the following (Note Problem 20):
(a) The alternative hypothesis as stated
(b) *Alternative hypothesis.* Proportion in favor is more than 0.80
(c) *Alternative hypothesis.* Proportion in favor is not 0.80

22. The level of significance of a test determines the size of the region of rejection. Suppose a test of a population proportion uses the 0.05 level of significance and the following alternative hypothesis: Proportion is less than 0.35. The significance level indicates that 0.05 (or 5%) of possible sample proportions fall in the lower-tail critical region if the population proportion is really 0.35. If the level of significance were changed to 0.02, then only 0.02 (or 2%) of possible sample proportions fall in the lower-tail critical region if the population proportion is really 0.35.
(a) Is the rejection region larger if the level of significance for a test of a population proportion is 0.02 or 0.01? Why?
(b) Is the critical region larger if the level of significance for a test of a population mean is 0.05 or 0.005? Why?

23. Refer to Problem 16. In this problem the level of significance is 0.05 and the critical value in the decision rule is 0.54. How would the critical value be affected if the level of significance were changed to 0.02? Why? (Note the discussion in Problem 22.)

24. In Problem 9 the significance level is 0.02 and the critical value is 7.0 ounces (as stated in the decision rule). After considering the discussion in Problem 22, how would the critical value be affected if—

(a) The level of significance were reduced to 0.01?

(b) The level of significance were increased to 0.05?

25. If a test of a population proportion has a two-sided alternative hypothesis, the region of rejection has two parts—a lower-tail region and an upper-tail region. If the level of significance is 0.05, then half of this (0.025) is assigned to each rejection region. In other words, the critical values are chosen so that 0.025 (or 2.5%) of the possible sample proportions fall in each rejection region if the null hypothesis is true. Refer to Problem 17. What is the probability that a type I error will be committed by selecting a random sample of adults where the sample proportion falls in the lower-tail part of the rejection region? The upper-tail part? In either part?

26. Refer to Problem 3 (Note Problem 25)—

(a) If the level of significance were reduced, how would this affect the lower-tail critical value? Why?

(b) If the level of significance were reduced, how would this affect the upper-tail critical value? Why?

Note 16A

Tests of Hypotheses Relating to a Population Proportion

This note discusses tests of hypotheses relating to a proportion where (1) the population is large, (2) sample size is less than 5% of the population, (3) population proportion is not too close to 0 or 1, and (4) sample size is at least 100 (a larger sample is usually preferred). Under these conditions, the sampling distribution of a proportion may be represented by the normal distribution, as discussed in **Note 13C**.

A soap manufacturer claims that 40% of the households in New York City prefer his soap product. A random sample of 500 households includes 185 who indicate such a preference. Conduct a test of the manufacturer's claim at the 0.05 level of significance, assuming that you feel his claim is too high.

The following hypotheses are formulated:

1. *Hypothesis to be tested.* Proportion of households who have the preference = 0.40
2. *Alternative hypothesis.* Proportion is less than 0.40

The sampling distribution of a proportion appropriate to this problem is represented by the normal distribution with a mean of 0.40 (as stated in the hypothesis to be tested). This is shown in Figure 16A.1. Since the level of significance for this test is 0.05 and the alternative hypothesis relates to proportions less than 0.40, we must find the proportion in the sampling distribution below which lie 5% of the possible sample proportions. This proportion is the critical value for this test and is determined as follows:

$$\text{Critical value} = z \times \sqrt{\frac{p \times (1 - p)}{n}} + p$$

where z = the value on the horizontal axis of the normal curve which cuts off 0.05 of the area under the curve in the lower-tail of the distribution

p = the population proportion stated in the hypothesis to be tested

n = sample size

Referring to Table B.1 in Appendix B, we determine $z = -1.64$ (as explained in **Note 13A**). This is shown in Figure 16A.1. Then, since $n = 500$ and $p = 0.40$, compute the following:

$$\text{Critical value} = -1.64 \times \sqrt{\frac{0.40 \times (1 - 0.40)}{500}} + 0.40$$
$$= 0.364$$

This critical value is shown in Figure 16A.1. Notice that the vertical line in the figure constructed at this critical value divides the area under the normal curve into two parts—a rejection region and an acceptance region.

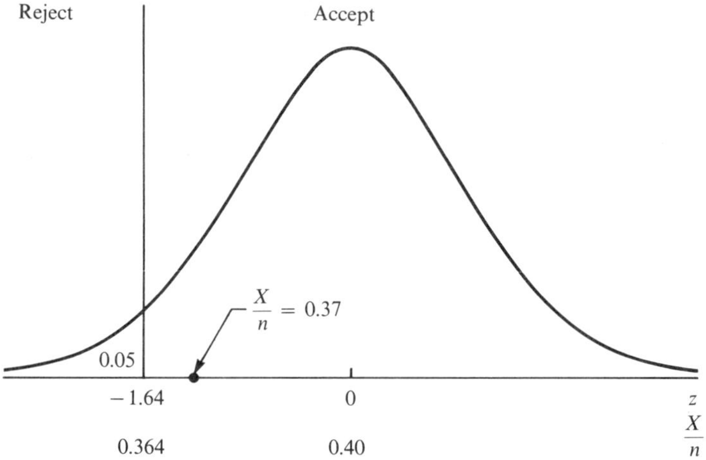

Figure 16A.1

Normal Distribution Approximating the Sampling Distribution of a Proportion, Showing the Regions of Acceptance and Rejection as Determined by the Critical Value $X/n = 0.364$

When a sample of size n is selected, we determine the number of sample observations (X) which fall into the category of interest and compute the sample proportion of such observations as X/n. For a particular sample, the sample proportion could have a value anywhere along the horizontal axis in Figure 16A.1. Since the alternative hypothesis indicates a concern that the claimed proportion 0.40 may be too high, and it is decided to take a risk of 0.05 that rejecting the claim will be an error (the significance level), Figure 16A.1 shows that if the proportion computed from the sample of households is less than 0.364, the manufacturer's claim (the null hypothesis) will be rejected. Otherwise, the claim will be accepted. This is why the lower-tail of the distribution in Figure 16A.1 is called the region of rejection and the balance of the distribution is called the region of acceptance.

Figure 16A.1 then, contains a decision rule which we can state in words as follows:

Decision rule If the proportion of households in the sample is less than 0.364, reject the null hypothesis, accept the alternative hypothesis, and do not accept the claim. Otherwise, accept the claim.

Based on the sample data, $X = 185$, the sample proportion which favors the soap product is $185/500 = 0.37$. Therefore, since 0.37 falls in the region of acceptance (as shown in the figure), accept the claim.

Of course, if the sample proportion were below 0.364 (and so would lie in the critical region), it would not *prove* that the claim is false. Such a sample proportion could occur even if the claim were true. However, it is *improbable* that such a sample proportion could be obtained (a probability of 0.05, the significance level) if the true population proportion is 0.40 as claimed.

Suppose the test is conducted to test the acceptability of the manufacturer's claim, with no feeling that the claim is either too high or too low. Then, the appropriate alternative hypothesis must be changed as follows:

Alternative hypothesis. Proportion is not 0.40 (it may be higher or lower)

When the alternative hypothesis is two-sided, the test must be constructed to provide protection against accepting the null hypothesis based on a sample proportion that is either too low or too high. In other words, a region of rejection must be provided in each tail of the sampling distribution. This is accomplished by first dividing the level of significance in half. (In our problem, $0.05 \div 2 = 0.025$.) Then, we find the proportion in the sampling distribution below which lies 0.025, or 2.5%, of the possible sample proportions and also the proportion above which lies 2.5% of the possible sample proportions. These proportions become the two critical values required for a test with a two-sided alternative.

Referring to Table B.1, we determine that $z = -1.96$ cuts off a lower tail with 2.5% of the total area under the normal distribution, and $z = 1.96$ cuts off 2.5% of the total area in the upper tail. Then, applying the equation for computing critical values we obtain the following:

$$\begin{pmatrix}\text{Lower-tail}\\ \text{critical value}\end{pmatrix} = -1.96 \times \sqrt{\frac{0.40 \times (1 - 0.40)}{500}} + 0.40 = 0.357$$

$$\begin{pmatrix}\text{Upper-tail}\\ \text{critical value}\end{pmatrix} = 1.96 \times \sqrt{\frac{0.40 \times (1 - 0.40)}{500}} + 0.40 = 0.443$$

Figure 16A.2 shows the normal curve approximation to the sampling distribution of a proportion, the z values, and critical values just determined. Notice that this figure shows two regions of rejection and a region of acceptance. This figure contains the following decision rule:

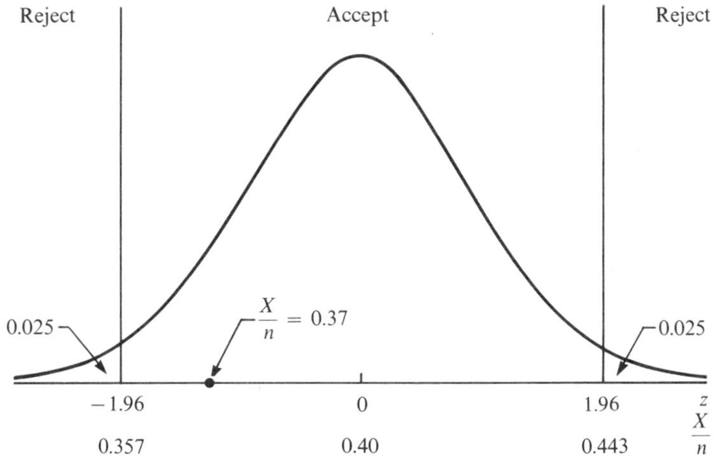

Figure 16A.2

Normal Distribution Approximating the Sampling Distribution of a Proportion, Showing the Region of Acceptance and Two Regions of Rejection as Determined by the Critical Values $X/n = 0.357$ and $X/n = 0.443$

Decision rule If the sample proportion is either less than 0.357 or more than 0.443, reject the null hypothesis, accept the alternative hypothesis, and do not accept the claim. Otherwise, accept the claim.

The sample proportion 0.37, as previously computed, lies in the region of acceptance (as shown in Figure 16A.2). Therefore, we accept the claim.

Note 16B

Tests of Hypotheses Relating to a Population Mean

This note discusses tests of hypotheses relating to a mean where (1) the population is large, (2) the sample includes 30 or more observations, and (3) the sample includes less than 5% of the population. Under these conditions, the sampling distribution of the mean may be represented by the normal distribution, as discussed in **Note 13B**. Determination of regions of rejection and acceptance follow the principles discussed in **Note 16A**. Critical values are computed as follows:

$$\text{Critical value} = z \times \frac{s}{\sqrt{n}} + \mu$$

where z = the value on the horizontal axis of the normal curve which cuts off an area in the lower-tail, upper-tail, or both tails combined (half in each tail), based on the level of significance of the test and the alternative hypothesis

s = sample standard deviation (computed as shown in **Note 9A**)

μ = population mean stated in the hypothesis to be tested

n = sample size

A sociologist claimed that if a social adjustment test he constructed were administered to the junior-high-school students in Chicago, they would obtain a mean score of 50. A random sample of 50 junior-high-school students obtained a mean score of 52 on this test, with a standard deviation of 12. Assuming you feel that the claim is an understatement, conduct a test of the sociologist's claim at the 1% significance level. The appropriate hypotheses are as follows:

1. *Hypothesis to be tested.* Population mean score = 50
2. *Alternative hypothesis.* Population mean is higher than 50

The alternative hypothesis is one-sided and relates to a possible mean above 50. Then, referring to Table B.1 in Appendix B, we determine that $z = 2.33$ cuts off an upper tail with 0.01 or 1% of the area under the normal curve. Applying the previous equation, the critical value is computed as follows ($s = 12$, $n = 50$, $\mu = 50$):

$$\text{Critical value} = 2.33 \times \frac{12}{\sqrt{50}} + 50 = 53.96$$

Decision rule If the sample mean score is more than 53.96, reject the null hypothesis, accept the alternative hypothesis, and do not accept the claim. Otherwise, accept the claim.

It is a good idea to sketch a normal curve and indicate the decision rule following the approach used in Figure 16A.1.

The sample mean score (52) falls in the region of acceptance. Therefore, we decide to accept the claim. If a test involves a two-sided alternative, the level of significance is divided in half and two regions of rejection and one region of acceptance are identified, as discussed in **Note 16A** and illustrated in Figure 16A.2.

Note 16C

The Chi-square Distribution

The chi-square distribution is a continuous distribution and is typically skewed to the right, as illustrated in Figure 16C.1. The variable of the distribution is χ^2 (the square of the Greek letter *chi*). This variable can take on the value 0 or any positive value, as indicated in the figure. Notice that the figure also presents the regions of rejection and acceptance for a one-tail test with the critical region in the upper tail.

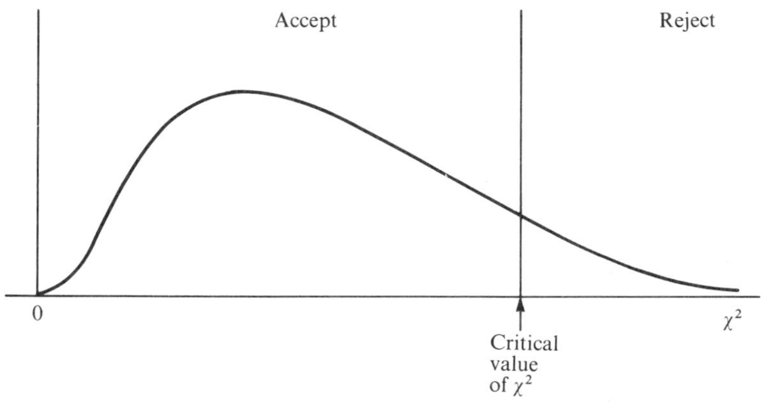

Figure 16C.1

A Chi-square Distribution

The shape of the χ^2 distribution pertinent to a problem depends on a technical consideration called "degrees of freedom" (usually abbreviated df). The number of df is computed differently for different types of problems. **Note 16D** shows how to determine the df for a test of hypothesis relating to the standard deviation.

Table B.2 in Appendix B presents critical χ^2 values for various levels of significance and for 1 to 30 df. The critical values in this table cut off upper-tail critical regions only. For example, the column headed ".05" presents χ^2 values for various df which cut off .05 of the area under the chi-square curve in the upper tail.

What about lower-tail critical regions? We know that for a 0.05 level of significance and a lower-tail critical region, the proportion of the area under the curve which lies in the lower tail is 0.05. Then 1 − 0.05, or 0.95, of the area lies to the right of the corresponding critical χ^2 value (the "upper tail"). In Table B.2, the column headed "0.95" contains the appropriate critical values for a test with a 0.05 level of significance and a lower-tail critical region.

Note 16D

Tests of Hypotheses Relating to the Standard Deviation of a Population

(Before studying this note, it is necessary to study **Note 16C**.)

This note discusses tests of hypotheses relating to the standard deviation of a population where the population distribution can be approximated by the normal distribution.

A hospital administrator claims that the number of cases treated in the emergency room during a weekend has a standard deviation of 10 cases. Test this claim using the 0.05 significance level and assuming that the claim is based on an honest and unprejudiced judgement of the emergency room caseload. (Assume that the number of cases treated in the emergency room is distributed approximately like the normal distribution.) A random sample of 18 weekends showed a standard deviation of 15 cases. The following hypotheses are formulated:

1. *Hypothesis to be tested.* Standard deviation = 10 cases
2. *Alternative hypothesis.* Standard deviation is not 10 cases (it may be higher or lower)

Critical values are computed as follows:

$$\text{Critical value} = \sqrt{\frac{\sigma^2 \times \chi^2}{n - 1}}$$

where σ^2 = population variance (square of standard deviation), assuming the null hypothesis is true

n = sample size

χ^2 = the chi-square value obtained from Table B.2 in Appendix B which cuts off an area under the χ^2 distribution curve in the lower tail or upper tail based on the level of significance of the test and the alternative hypothesis.

The number of degrees of freedom for the type of tests we are considering in this note is computed as $n - 1$.

According to the null hypothesis in our problem, the standard deviation in the population is 10, so $\sigma^2 = 10 \times 10 = 100$. Since the alternative hypothesis is two-sided, we must determine two critical values, a lower-tail critical value which cuts off $1/2 \times 0.05$ (level of significance of the test) $= 0.025$ of the area under the chi-square distribution in the lower tail and an upper-tail critical value which cuts off 0.025 of the area in the upper tail, as explained in **Note 16C**.

Since $n = 18$ in our problem, $\text{df} = 18 - 1 = 17$. Then, for the lower-tail critical value, we find the χ^2 value in the column headed "0.975" (since $1 - 0.025 = 0.975$) in Table B.2 for 17 df. This χ^2 value is 7.564. For the upper-tail critical value, we find the χ^2 value in the column headed "0.025" for 17 df. This χ^2 value is 30.191. Applying the equation for critical values, we obtain the following:

$$\begin{pmatrix}\text{Lower-tail} \\ \text{critical value}\end{pmatrix} = \sqrt{\frac{100 \times 7.564}{18 - 1}} = 6.7$$

$$\begin{pmatrix}\text{Upper-tail} \\ \text{critical value}\end{pmatrix} = \sqrt{\frac{100 \times 30.191}{18 - 1}} = 13.3$$

Decision rule If the sample standard deviation is either less than 6.7 cases or more than 13.3 cases, reject the null hypothesis, accept the alternative hypothesis, and conclude that the standard deviation of the number of cases treated in the emergency room during a weekend is not equal to 10. Otherwise, conclude that the standard deviation $= 10$.

The sample standard deviation (15) exceeds the upper-tail critical value. Hence, according to the decision rule, we conclude that the standard deviation does not equal 10.

17

Decision Making: Comparisons

Many decision-making problems involve a *comparison*. For example, a manufacturer may wish to compare the productivity of men and women by comparing the average number of units produced per hour by men and women workers. This is an illustration involving the *comparison of population means*. A sociologist may wish to compare the proportions of newly married couples who purchase a home during the first year of marriage in two cities. This illustrates the *comparison of population proportions*. Or, an education specialist may wish to compare the variation in achievement test scores when a foreign language is taught to high-school students by two different methods. This illustrates the *comparison of population standard deviations* (*or population variances*).

In this chapter we will study the comparison of parameter values (measures) in *two populations*. Statistical tests for comparisons are based on data obtained from *two samples*, one selected from each population. These samples may be *independent* or *dependent*. Two samples are *independent* if each is selected on a random basis from its population, so that selection for one sample in no way influences selection for the other sample. Two samples are *dependent* if there is a tie-in in the selection processes for the samples.

Dependent samples may be obtained in different ways. For example, the effectiveness of a reducing diet may be tested by selecting a random sample of subjects and determining the weight of each subject *before* and *after* the dieting period. The "before" and "after" weights represent a pair of dependent weight samples, since there is a weight in each sample for each subject. Notice that this illustration involves a *single population of subjects.* However, the statistical test is based on *two populations of weights*—the population of weights of all subjects before the diet and the population of "after diet" weights if all the subjects were to participate in the diet plan. Notice that the population of "after diet" weights is actually a *nonexistent population.*

A sociologist may wish to test whether degree of sexual activity among teenage girls is family related. He could select a sample of families with at least two unmarried teenage daughters per family and randomly assign one daughter to each sample. The girls would be asked how many times they had sexual relations. This would result in a pair of dependent samples of replies because there is a reply in each sample for a given family.

Sometimes dependent samples are obtained by *matching.* Suppose an agricultural researcher wishes to compare the relative effectiveness of two methods of planting soybeans. He could select a sample of plots for the trial of one method and a second sample of plots for the trial of the other method. However, the plots could be selected to *pair off* (*match*) on some relevant characteristic, such as soil acidity. Then, if a plot in one sample has a given acidity measure, a plot in the other sample will have the same acidity measure. Sometimes sample items are selected to pair off on two or more characteristics. These are called *matched samples.* When matched samples are used, it is more likely that observed differences between the two samples are caused by the *experimental variable* (in our illustration, planting method) and not by *extraneous factors* (such as soil acidity).

17.1 Comparing Proportions

A department store has its main outlet in the downtown shopping area and a large branch store in a suburban location. The general manager claims that the proportion of customers who return all or part of their purchases is the same in both locations. He asks his accountant to select a sample of customers at each location to test this claim. The accountant finds that in a random sample of 350 customers in the downtown store, 52 returned some or all of their purchases. A random sample of 300 customers in the suburban store included 41 who returned some or all of their purchases. Should the accountant conclude that the proportion of customers who return some or all of their purchases is the same for both stores, using the 0.05 significance level?

1. *Hypothesis to be tested.* The proportion is the same in both stores
2. *Alternative hypothesis.* The proportion is not the same

Decision rule If the larger sample proportion minus the smaller sample proportion is greater than 0.053, reject the null hypothesis, accept the alternative hypothesis, and conclude that the proportions are not the same. Otherwise, conclude that they are the same.

This decision rule indicates that even if the proportion is the same in both stores, the probability is 0.05 (the level of significance) that the difference between proportions could be more than 0.053 (either way) for a particular pair of samples. A probability of 0.05 indicates that it is very unlikely that such a difference will occur if the nu hypothesis is true (5 chances in 100). Therefore, if such a difference occurs, it lead to rejection of the null hypothesis.

According to sample results, the proportion of customers making returns is 52/350 = 0.149 for the downtown store and 41/300 = 0.137 for the suburban store. The difference between these proportions is 0.149 − 0.137 = 0.012, which falls in the region of acceptance. Hence, it is concluded that the proportions are the same for the two stores.

(**Note 17A** at the end of the chapter shows how to construct a test of hypothesis relating to the comparison of population proportions.)

17.2 Comparing Means

The personnel manager of a large telephone company believes that male and female operators are equally productive and decides to test this. He selects a random sample of 75 male operators from his large operator staff and determines that these operators handled an average (mean) of 21.6 calls per hour the previous week, with a standard deviation of 8.9 calls. A random sample of 80 female operators is found to have handled a mean of 18.2 calls per hour during the same period with a standard deviation of 10.1 calls. Using the 0.01 level of significance, would you accept or reject the feeling of the personnel manager?

1. *Hypothesis to be tested.* The mean number of calls handled by male and female operators is the same
2. *Alternative hypothesis.* The mean is not the same

Decision rule If the larger sample mean minus the smaller sample mean is greater than 3.95, reject the null hypothesis, accept the alternative hypothesis, and conclude that the two means are not the same. Otherwise, conclude that the means are the same.

Applying the sample results, we obtain 21.6 − 18.2 = 3.4. This difference is less than the critical value (3.95). Consequently, conclude that the mean number of calls handled per hour is the same for male and female telephone operators.

The foregoing problem illustrates a test of an hypothesis comparing two population means based on *independent random samples*. (**Note 17B** at the end of the chapter shows how to construct such a test.)

Are first-grade boys and girls equal in their ability to learn how to read? A random sample of 44 first-grade boys were selected and a *matched sample* of 44 first-grade girls were selected. They were matched on certain family characteristics (educational attainment of the parents, income level, etc.). The two samples of first graders were put through a similar program of reading instruction and practice. Then a reading proficiency test was administered to each sample and the 44 scores in one sample paired with those in the other sample. That is, the scores obtained by a boy and girl who matched on family characteristics were paired. This provided 44 pairs of scores.

The difference between the scores in each pair was obtained (boy score minus girl score). The mean of the 44 score differences was computed and found to be 0.36 point with a standard deviation of 4.70 points. Could it be concluded, at the 0.05 level of significance, that first-grade boys and girls are equal in their ability to learn how to read?

1. *Hypothesis to be tested.* The mean score on the reading proficiency test is the same for first-grade boys and girls
2. *Alternative hypothesis.* The mean score is not the same

Decision rule If the mean of the score pair differences is either less than -1.39 or more than 1.39, reject the null hypothesis, accept the alternative hypothesis, and conclude that the mean test score is not the same for first-grade boys and girls. Otherwise, conclude that the mean is the same.

The mean difference computed for the sample of 44 pairs of scores (0.36 point) falls between the two critical values (-1.39 and 1.39). Therefore, conclude that the mean test score is the same for first-grade boys and girls and that they are equal in ability to learn how to read.

The foregoing problem illustrates a test of hypothesis comparing two population means based on *dependent samples*. (**Note 17C** at the end of the chapter shows how to construct such a test.)

17.3 Comparing Variances

A paper-cutting machine in a factory required special care in adjustment if the length of sheets cut were to vary little from cut to cut. A random sample of 20 cuttings made after one worker adjusted the machine showed a standard deviation of cut lengths of 0.05 centimeter. A random sample of 25 cuttings made after another worker adjusted the machine showed a standard deviation of cut lengths of 0.07 centimeter. Would you conclude that the variation in cut lengths is the same when the machine is adjusted by either worker? Use the 0.10 significance level for the test.

1. *Hypothesis to be tested.* The variance (or standard deviation) is the same for cut length if the machine is set by either worker
2. *Alternative hypothesis.* The variance is not the same

Decision rule If the ratio of the larger sample variance (the square of the standard deviation) to the smaller is greater than 2.11, reject the null hypothesis, accept the alternative hypothesis, and conclude that the variance is not the same for the two workers. Otherwise, conclude that it is the same.

Notice that in tests comparing two variances, the critical value is a *ratio* (2.11). The decision rule indicates that even if the variance of cuts is the same for the two workers, the probability is 0.10 (the significance level) that the *variance ratio* for a particular pair of samples could exceed 2.11. A probability of 0.10 indicates that it is unlikely that such a ratio of sample variances will be obtained if the null hypothesis is true. Therefore, if such a variance ratio is obtained, it leads us to rejection of the null hypothesis.

The sample variances are $0.05 \times 0.05 = 0.0025$ for the first worker and $0.07 \times 0.07 = 0.0049$ for the second. The ratio of the variances is $0.0049/0.0025 = 1.96$, which is less than the critical value. Therefore, conclude that the variance is the same for the two workers.

(The decision rule in the foregoing problem is based on a new concept, the F statistic, which is related to a theoretical distribution called the F distribution. **Note 17D** at the end of chapter discusses this distribution and the general approach for computing the F statistic. **Note 17E** shows how to construct a test of hypothesis relating to the comparison of two population variances.)

Review

Concepts and Terminology

Dependent samples
Independent samples
Matched samples
Variance ratio

Problems

1. Indicate whether each of the following pairs of samples are independent or dependent and why.
   (a) Test scores on a business management exam and test scores on an accounting principles exam for a sample of college freshmen
   (b) A random sample of 18 productivity rates of experienced employees selected from firm A and a random sample of 18 productivity rates of experienced employees selected from firm B
   (c) A sample of typing speeds selected from typists using typewriter A and a sample of typing speeds selected from typists using typewriter B (The two samples were matched on months of experience and age.)
   (d) Time reactions to an electric shock by a sample of subjects before administration of a drug and time reactions of these subjects after the drug was administered

2. In Problem 1, part (d), which of the two samples represents a sample from a nonexistent population? Why?

3. A manufacturer uses a certain small part in production. He buys this part from firm A and firm B. He wants to verify that the mean weight per part is the same for both firms, even though he feels that they are actually the same. He plans to select a random sample of parts from the last two deliveries of each firm and conduct a test to compare the mean weights. He plans to conduct the test with a probability of 0.01 that he will be in error if he concludes that the mean weights are not the same.

(a) Formulate the hypothesis to be tested and the appropriate alternative hypothesis.
(b) What is the level of significance of this test?
(c) What is the appropriate alternative hypothesis if he actually believes that the mean weight for the parts is less for firm B?

4. A large law firm wishes to determine whether the mean number of days required per case was the same for cases handled during the last two years as for cases handled the previous two years. If a statistical test is to be conducted to compare the mean number of days for the two periods, formulate the following:
(a) The appropriate hypothesis to be tested
(b) The alternative hypothesis if it is felt that the last two years has a larger mean than the previous two years
(c) The alternative hypothesis if the firm believes that the two means are the same

5. A large food store chain is considering the purchase of a carload of apples and has to choose between two separate carloads, one from orchard A and one from orchard B. If a statistical test is to be conducted to compare the proportion of apples in each carload that are infected by apple scab, formulate the following:
(a) The null hypothesis for this test
(b) The alternative hypothesis if there is no feeling that the proportions are different

6. A department store plans to conduct a test to verify that the proportion of returns is the same for sales made by male clerks and female clerks. Formulate the appropriate hypotheses for conducting the statistical test.

7. It is desired to test whether the same proportion of pupils in two school districts come from homes with divorced parents. A random sample of 550 pupils in school district I includes 180 who come from such homes and a random sample of 500 pupils in school district II includes 150 who come from such homes. Could it be concluded that the proportions for the two populations of pupils are the same? The level of significance for this test is set at 0.05. The decision rule is constructed as follows: If the larger sample proportion minus the smaller exceeds 0.057, reject the null hypothesis (that these proportions are equal for the two populations) and accept the alternative hypothesis (that they are not equal). Otherwise, conclude that the proportions are equal. Apply the sample results and make the decision.

8. It is desired to test whether the mean price of a commodity is the same in two regions. A random sample of 40 stores in region A shows a mean price of 24¢ per pound (with a standard deviation of 8¢) and a random sample of 45 stores in region B shows a mean price of 22¢ per pound (with a standard deviation of 9¢). The level of significance for this test is 0.02. The decision rule is formulated as follows: If the larger sample mean minus the smaller exceeds 4.29¢ per pound, reject the null hypothesis (that the mean price is the same in the two regions) and accept the alternative hypothesis that they are not the same. Otherwise, conclude that the mean prices in the regions are the same. Apply the sample results and make the decision.

9. The standard deviation for a sample of 21 test A scores randomly selected from the records of a boys' school is 2.1 points. The standard deviation is 1.8 points for a sample of 16 test A scores randomly selected from the records of a girls' school. A statistical test is to be conducted at the 0.05 significance level. The following decision rule is to be applied: If the ratio of the larger sample variance to the smaller exceeds 2.33, reject the null hypothesis (that the two population variances are equal) and accept the alternative hypothesis that the variances are not equal. Otherwise, conclude that they are equal. Apply the sample findings and make the decision.

Note 17A

Tests of Hypotheses for the Comparison of Two Population Proportions

This note applies to tests where the following conditions are met by each population and each sample: (1) population is large, (2) sample size is less than 5% of the population, (3) population proportion is not too close to 0 or 1, and (4) sample size is at least 100.

The tests discussed in this note are based on two independent random samples, one sample from each population. Let us consider the problem discussed in Section 17.1 relating to a department store with a main outlet downtown and a suburban branch. This problem involved a sample of 250 customers from the downtown store which included 52 customers who returned some or all of their purchases and a sample of 300 customers from the suburban store which included 41 such customers. The proportion of customers making returns is $52/250 = 0.149$ for the downtown-store sample and $41/300 = 0.137$ for the suburban-store sample. The difference between sample proportions, $0.149 - 0.137 = 0.012$, is the statistic used in conducting the test.

The hypotheses stated in Section 17.1 may be written somewhat more technically as follows:

1. *Hypothesis to be tested.* The difference between the population proportions for the two store outlets = 0 (or, the proportions are equal)
2. *Alternative hypothesis.* The difference is not equal to zero (or, the proportions are not equal)

The decision rule is based on the *sampling distribution of the difference between sample proportions*, assuming that the samples are independently selected. According to statistical theory, this sampling distribution is approximately like the normal distribution, so the normal curve may be used for constructing the decision rule. Critical values are computed as follows:

$$\text{Critical value} = z \times \sqrt{\frac{X}{n} \times \left(1 - \frac{X}{n}\right) \times \left(\frac{1}{n_1} + \frac{1}{n_2}\right)}$$

where z = the value on the horizontal axis of the normal curve which cuts off an area in the lower tail, upper tail, or both tails combined (half in each tail) based on the level of significance of the test and the alternative hypothesis

n_1 = size of the first sample

n_2 = size of the second sample

$$\frac{X}{n} = \frac{X_1 + X_2}{n_1 + n_2}$$

where X_1 = number of the n_1 sample observations in the category of interest and X_2 = number of the n_2 sample observations in the category of interest. Therefore, X/n = the proportion of observations in both samples combined that are in the category of interest.

Returning to our problem, we have $n_1 = 350$, $X_1 = 52$, $n_2 = 300$, and $X_2 = 41$ so that

$$\frac{X}{n} = \frac{52 + 41}{350 + 300} = 0.143$$

Since the alternative hypothesis is two-sided, we must determine two critical values, a lower-tail critical value which cuts off $1/2 \times 0.05$ (the level of significance of the test as noted in Section 17.1) $= 0.025$ of the area under the normal curve in the lower tail and an upper-tail critical value which cuts off 0.025 of the area in the upper tail. Referring to Table B.1, we determine that $z = -1.96$ (in the lower tail) and $z = 1.96$ (in the upper tail) cut off 0.025 of the normal curve area in each tail. Then, these are the z values to use to compute the critical values for the test.

Substituting the foregoing values for X/n and z into the critical value equation, we obtain

$$\begin{pmatrix}\text{Lower-tail} \\ \text{critical value}\end{pmatrix} = -1.96 \times \sqrt{0.143 \times (1 - 0.143) \times \left(\frac{1}{350} + \frac{1}{300}\right)} = -0.053$$

$$\begin{pmatrix}\text{Upper-tail} \\ \text{critical value}\end{pmatrix} = 1.96 \times \sqrt{0.143 \times (1 - 0.143) \times \left(\frac{1}{350} + \frac{1}{300}\right)} = 0.053$$

The decision rule (stated somewhat more technically than in Section 17.1) is: If the difference between sample proportions is either less than -0.053 or more than 0.053, reject the hypothesis tested, accept the alternative hypothesis, and conclude that the proportion of customers making returns is not the same in both outlets. Otherwise, conclude that this proportion is the same.

Since the difference between sample proportions in our problem (0.012) lies between the two critical values, it falls in the region of acceptance. Therefore, conclude that the population proportions are the same for the two outlets.

Note 17B

Tests of Hypotheses for the Comparison of Two Population Means (Independent Samples)

This note applies where independent random samples are selected, one from each population, and the following conditions are met by each population and each sample: (1) population is large, (2) sample includes 30 or more observations, (3) sample size is less than 5% of the population.

Let us consider the problem relating to the telephone company operators discussed in Section 17.2. The difference between the sample means, 21.6 (male operators) − 18.2 (female operators) = 3.4 calls, is the statistic used in conducting the test. The hypotheses stated in Section 17.2 may be expressed somewhat more technically as follows:

1. *Hypothesis to be tested.* The difference between the population means for male and female operators = 0 (or, the means are equal)
2. *Alternative hypothesis.* The difference is not equal to zero (or, the means are not equal)

The decision rule is based on the *sampling distribution of the difference between sample means* which is similar to the normal distribution, so the normal curve can be used for constructing the decision rule. Critical values are computed as follows:

$$\text{Critical value} = z \times \sqrt{\frac{s_1^2}{n_1} + \frac{s_2^2}{n_2}}$$

where z = the value on the horizontal axis of the normal curve which cuts off an area in the critical region or in both critical regions combined in a two-tail test (half in each tail) based on the significance level of the test and the alternative hypothesis

n_1 = size of the first sample

n_2 = size of the second sample

$s_1^2 = s_1 \times s_1$(where s_1 is the standard deviation of the n_1 sample)

$s_2^2 = s_2 \times s_2$ (where s_2 is the standard deviation of the n_2 sample)

In our illustrative problem, the alternative hypothesis is two-sided, so two critical values must be determined. Referring to Table B.1, we determine that $z = -2.58$ (in the lower tail) and $z = 2.58$ (in the upper tail) cut off 0.005 of the normal curve area in each tail. Then, since the significance level of the test is 0.01 (as noted in Section 17.2), these are the z values to use in computing the critical values for the test.

As noted in Section 17.2, $s_1 = 8.9$, $s_2 = 10.1$, $n_1 = 75$, and $n_2 = 80$. Substituting these values and the z values just determined into the equation for critical value, we obtain the following:

$$\begin{pmatrix}\text{Lower-tail} \\ \text{critical value}\end{pmatrix} = -2.58 \times \sqrt{\frac{(8.9)^2}{75} + \frac{(10.1)^2}{80}} = -3.95$$

$$\begin{pmatrix}\text{Upper-tail} \\ \text{critical value}\end{pmatrix} = 2.58 \times \sqrt{\frac{(8.9)^2}{75} + \frac{(10.1)^2}{80}} = 3.95$$

The decision rule (stated somewhat more technically than in Section 17.2) is: If the difference between sample means is either less than -3.95 calls or more than 3.95 calls, reject the hypothesis tested, accept the alternative hypothesis, and conclude that the mean number of calls handled by male and female operators are not the same. Otherwise, conclude that the means are the same.

Since the difference between sample means (3.4 calls) lies between the two critical values, it falls in the region of acceptance. Therefore, we conclude that the mean number of calls handled per hour is the same for male and female telephone operators.

Note 17C

Tests of Hypotheses for the Comparison of Two Population Means (Dependent Samples)

The second illustrative problem in Section 17.2 involves matched samples of first-grade boys and girls which represent samples from two populations—a population of first-grade boys and a population of first-grade girls. Think of the two populations of first-graders as being matched (paired) on the same basis as the samples of 44 boys and girls. Imagine that all the boys and girls in the two populations were put through the program of reading instruction and practice and took the reading proficiency test. Then, if the difference between scores in each pair were obtained, we would have a population of score differences.

When a test of an hypothesis comparing two population means involves dependent samples, the test is conducted based on the *population of pair differences*. In other words, the test comparing the means of two populations becomes a test of the mean of a population of differences. Then, the test is conducted as described in **Note 16B**.

The hypotheses stated in Section 17.2 may be expressed somewhat more technically as follows:

1. *Hypothesis to be tested.* The difference between the population mean scores for first-grade boys and girls = 0 (or, the means are equal)
2. *Alternative hypothesis.* The difference is not equal to zero (or, the means are not equal)

Critical values are computed according to the equation in **Note 16B**. However, since the null hypothesis states that the difference between population means is equal to zero, μ (the mean for the *population of pair differences*) = 0. Therefore, we may write the equation for critical values as follows:

$$\text{Critical value} = z \times \frac{s}{\sqrt{n}}$$

where z = (as in **Note 16B**)
s = standard deviation computed for the sample of *pair differences*
n = number of sample *pairs*

As noted in Section 17.2, the level of significance is 0.05 and since the alternative hypothesis is two-sided, $z = -1.96$ and $z = 1.96$ are to be used to compute critical values. According to Section 17.2, $n = 44$ and $s = 4.70$ points. We compute critical values as follows:

$$\begin{pmatrix}\text{Lower-tail}\\ \text{critical value}\end{pmatrix} = -1.96 \times \frac{4.70}{\sqrt{44}} = -1.39$$

$$\begin{pmatrix}\text{Upper-tail}\\ \text{critical value}\end{pmatrix} = 1.96 \times \frac{4.70}{\sqrt{44}} = 1.39$$

The decision rule is stated as in Section 17.2 and the same conclusion is drawn.

Note 17D

The *F* Distribution and Computing the *F* Statistic

Figure 17D.1 presents an F distribution and indicates that this distribution is skewed to the right, continuous, and that F can take on any value from 0 to infinity. The figure also shows the regions of acceptance and rejection for a one-tail test, with the critical region in the upper tail.

Generally, the F statistic is computed as the ratio of two variances (variance 1/variance 2), where each variance is computed from sample data and the samples are randomly selected from populations which have equal variances and are normally distributed. (Recall from Chapter 9 that the square of the standard deviation is called the variance.)

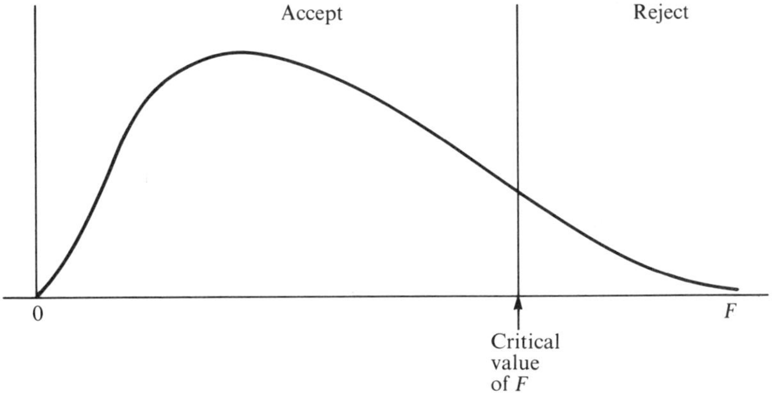

Figure 17D.1

An F Distribution

The F distribution pertinent to a problem is determined by two numbers of "degrees of freedom" (abbreviated as df)—df related to the numerator variance and df related to the denominator variance. **Note 17E** shows how to determine the numbers of df for a test relating to the comparison of two variances.

Tables B.3 and B.4 in Appendix B present critical F values for various pairs of df. The critical values in these tables cut off upper-tail critical regions. The F values in Table B.3 cut off 0.05 of the total area under the curve in the upper tail for various pairs of df and the F values in Table B.4 cut off 0.01 of the total area under the curve in the upper tail.

In statistical tests where a lower-tail critical region may be involved, compute the F statistic by placing the *larger* of the two sample variances in the numerator. Then, the test may always be conducted with the critical region in the upper tail of the distribution.

Note 17E

Tests of Hypotheses for the Comparison of Two Population Variances

(Before studying this note, it is necessary to study **Note 17D**.)

This note applies where independent random samples are selected, one from each population, and population distributions are approximately like the normal distribution.

Refer to the problem in Section 17.3 relating to a cutting machine. Assume that cut lengths are distributed like the normal curve. The null hypothesis in the test states that the variance of cut lengths is equal for the two workers. When two normally distributed populations have equal variances, as in our illustration, the sampling distribution of the F statistic is like the F distribution (Figure 17D.1), where the F statistic is computed as follows: variance computed for a sample selected from one normal population divided by the variance computed for a sample selected from the other normal population.

Since the alternative hypothesis is two-sided, we must compute a lower-tail critical value and an upper-tail critical value. As indicated in **Note 17D**, Tables B.3 and B.4 present critical F values for upper-tail regions of rejection only. Therefore, compute the F statistic as follows:

$$F = \frac{\text{larger sample variance}}{\text{smaller sample variance}}$$

Then, only the upper-tail critical value need be computed. Let us consider why this is so.

Notice that the F statistic is computed as a ratio. If the null hypothesis is true so that the two population variances are equal, the ratio of the variances will equal unity. However, the ratio of the variances for two samples, one sample from each population, may well differ considerably from unity. If the numerator variance is smaller than the denominator variance, the F statistic will be less than unity and the lower-tail critical region will determine when to reject the null hypothesis. On the other hand, if the numerator variance is larger, the F statistic will be greater than unity and the upper-tail critical region will determine when to reject the null hypothesis.

The size of the region of rejection is determined by the level of significance of the test. In the problem we are considering, the level of significance is 0.10, as noted in Section 17.3. Then half of this, 0.05, is the size of the critical region in each tail of a two-tail test (as in our problem). If we follow the practice of always putting the larger sample variance in the numerator, we will need only to compute the upper-tail critical value and we will conveniently be able to use Tables B.3 and B.4.

As shown in Section 17.3, the standard deviations for the samples are 0.05 centimeter and 0.07 centimeter; so the sample variances are $0.05 \times 0.05 = 0.0025$ and $0.07 \times 0.07 = 0.0049$. Then, the F statistic is $0.0049/0.0025 = 1.96$, as shown in Section 17.3.

How do we determine the upper-tail critical value? The number of df is determined for the numerator as (sample size -1). In our problem, the 0.0049 variance was computed from a sample of 25 cuttings, so the numerator df $= 25 - 1 = 24$. Similarly, the denominator variance 0.0025 was computed from a sample of 20 cuttings, so that the denominator df $= 20 - 1 = 19$. Referring to Table B.3, read out $F = 2.11$ for 24 df (numerator) and 19 df (denominator). Then, the decision rule is stated as in Section 17.3 and the conclusion is reached as shown.

18

Decision Making: Further Topics

The decision-making procedures presented in chapters 15, 16 and 17 involved three steps:

1. Formulation of the hypothesis to be tested and the alternative hypothesis relating to the decision to be made in nontechnical terms
2. Statement of the decision rule in concrete terms
3. Application of the decision rule to sample data and making the decision

In this chapter we will consider some widely used statistical tests in which, because of technical reasons, the decision rule cannot be stated in nontechnical, concrete terms. However, the appropriate hypotheses can be formulated in nontechnical terms and the evaluation of sample data in terms of the decision rule can be followed by the nonstatistician quite easily.

The decision rule will be expressed in terms of one of two concepts—the *chi-square statistic* or the *F statistic*. These statistics represent theoretical probability distributions that are too technical to explain in nonstatistical terms. However, the general framework of the decision-making procedure will not be difficult to follow.

The decision-making procedures presented which use a chi-square statistic in the decision rule are the *chi-square tests* frequently used by statisticians. The decision-making procedure presented which uses the *F* statistic is a frequently used *analysis of variance* test. Many nonstatisticians whose work involves the use of statistical decision-making procedures will no doubt hear statisticians speak about these statistical tests. (**Note 16C** at the end of chapter 16 described the chi-square distribution and **Note 18A** at the end of this chapter shows how the chi-square statistic is computed for the chi-square tests in this chapter. **Note 17D** at the end of chapter 17 describes the *F* distribution and indicates the general approach for computing the *F* statistic).

18.1 Comparing Several Proportions

There has been considerable discussion about the desirability of abortions for those women who wish them. A public opinion research organization is interested in determining the response of women to the question "Do you favor abortion if it is required that it be performed at a sufficiently early stage of the pregnancy, in a hospital, and by a licensed physician?" A random sample of 183 women were asked this question in city A, 137 women in city B, and 206 women in city C. The responses are summarized in Table 18.1. Based on these data could it be concluded that the proportions responding yes are the same for the three cities? Use the 0.05 level of significance.

Table 18.1

Responses of Three Samples of Women to a Question on Abortion

Response	City A Sample	City B Sample	City C Sample	Total
Yes	107	70	89	266
No	76	67	117	260
Total	183	137	206	526

In our illustration we are dealing with three populations—women in city A, women in city B, women in city C. In each population and in each sample there are two response categories, the "Yes" category and the "No" category. The category of interest in this survey is the "Yes" category. The statistical test will compare the proportions of yes responses in the three samples.

Notice from Table 18.1, last column, that 266 women out of the 526 in the three samples combined responded yes. In other words, the proportion of yes responses for the three samples combined is $266/526 = 0.51$. If the proportion of yes responses is actually the same in the three cities, then 0.51 represents an estimate of this proportion for each city. However, according to the sample data, the proportion of yes responses is $107/183 = 0.58$ in city A, $70/137 = 0.51$ in city B, and $89/206 = 0.43$ in city C. Could the proportions in city A and city C samples represent only *chance departures* from 0.51 due to the vagaries of random sampling? Or, are these proportions different enough from 0.51 to indicate that the proportion of yes responses is not the same in the three cities? The following hypotheses are formulated:

1. *Hypothesis to be tested.* The proportions in the three cities are the same
2. *Alternative hypothesis.* The proportions are not the same (at least one is different)

Decision rule If the chi-square statistic computed from the sample data is greater than 5.991 reject the hypothesis tested, accept the alternative hypothesis, and conclude that the proportions are not the same. Otherwise, conclude that the proportions are the same.

The chi-square statistic computed from the sample data is 9.097, which exceeds the critical value (5.991). Therefore, conclude that the proportions in the three cities are not the same.

(**Note 18B** at the end of the chapter shows how to construct a chi-square test to compare several population proportions.)

18.2 Tests of Independence

Statistical tests which seek to evaluate whether or not two characteristics are in some way related are called *tests of independence*. For example, in the previous illustration concerning abortion, do younger women tend to answer yes more often or less often than older women? In other words, are the two characteristics "response to the question" and "age" *dependent* (*related*) or *independent*? Or, is response to the question related to whether or not the women are married? That is, are the characteristics "response to the question" and "marital status" dependent or independent?

As another illustration, a department store manager may wish to determine whether "preference for credit purchases" and "sex" are independent. That is, do men and women have similar preferences concerning buying on credit or paying cash?

A large insurance firm is planning a sales expansion program. A question bothering the vice-president of sales is whether he should take on as new agents only those with at least some college education. He thinks that an agent's sales volume may be related to his level of education. He asks his statistician to look into this question. The statistician selects a random sample of 350 agents from the firm's large force and classifies them according to educational attainment and sales volume last year (level I represents the lowest sales volume). The sample results are summarized in Table 18.2. Using the 0.01 significance level, should the statistician conclude that "educational attainment" and "sales volume" are independent?

Table 18.2

A Sample of 350 Insurance Agents Classified by Sales Volume and Educational Attainment

	Educational Attainment			
Sales Volume	College Graduate	Some College	High-school Graduate	Total
Level I	15	22	30	67
Level II	25	31	45	101
Level III	30	43	20	93
Level IV	40	35	14	89
Total	110	131	109	350

Table 18.2 is a *two-way classification frequency table* since it presents the frequencies (counts of agents) corresponding to the *joint categories* of two classifying characteristics (variables), sales volume and educational level. Statisticians call this a *contingency table.*

Notice from the contingency table, last column, that 67 out of the sample of 350 agents fall in the level I sales volume, so that 67/350 = 0.1914 or 19.14% of the sample of agents are in this sales volume category. As discussed in chapter 10, this indicates that the probability is about 0.19 (19 chances in 100) that an agent in the population falls in the level I sales volume class. According to the total row in Table 18.2, 110 out of the 350 agents are college graduates; so 110/350 = 0.3143 or 31.43% of the sample of agents are college graduates. This indicates that the probability is about 0.31 that an agent in the population is a college graduate.

Combining the foregoing information, statisticians would determine that 0.1914 × 0.3143 = 0.0602 or 6.02% of the agents in the sample are *both* college graduates *and* produce at the level I volume of sales. Hence, the probability is about 0.06 (6 chances in 100) that an agent in the population is both a college graduate and produces level I sales volume. Multiply this probability by the total number of agents in the sample (350), to obtain 0.0602 × 350 = 21.1. This indicates that the *expected number of agents* in a sample of 350 who are both college graduates and produce level I volume of sales is 21.1 agents, on the average. However, according to the contingency table, 15 agents in the sample fall in this category.

Following this approach we can compute the *expected number of agents* who should fall in each of the 12 *joint-classification categories* or *cells* in the contingency table. For example, notice that 101 of the 350 agents in the sample produced at the level II volume of sales and 109 of the 350 agents are high-school graduates. Then, 101/350 = 0.2886, 109/350 = 0.3114, and 0.2886 × 0.3114 = 0.0899. Finally, 0.0899 × 350 = 31.5 represents the expected number of the 350 agents who should fall in the joint-classification cell "high-school graduate and sales volume level II." Actually, 45 agents fall into this cell in the contingency table.

Table 18.3 presents the expected number of agents in the sample who should fall in each of the 12 categories (or cells) in parentheses, the actual number which did fall in each category (called the *observed number of agents*), and the difference between the expected and observed numbers in each cell. The reader may be wondering what is meant by *expected number* (or *expected frequency*). These frequencies indicate the number of agents expected in each cell *if actually there is no relationship between sales volume and educational attainment.* In other words, these are the numbers of agents to be expected if the two characteristics "sales volume" and "educational attainment" are *independent.*

Notice in Table 18.3 that the expected frequencies and observed frequencies differ. What does this mean? Actually, even if sales volume is not related to educational level, it is not to be expected that the sample cell frequencies (the observed frequencies) will exactly equal the expected frequencies. Differences due to the vagaries of random sampling are to be expected. However, are these differences too large to be due merely to the *chance effects* of random sampling? That is, are these differences large enough to indicate that the two characteristics are not independent, but in some way related (dependent). These differences provide the basis for constructing the test of independence. The appropriate hypotheses are as follows:

Table 18.3

Observed and Expected Frequencies for Contingency Table in Table 18.2 *and the Differences between Them (Expected Frequencies in Parentheses)*

	Educational Attainment		
Sales Volume	College Graduate	Some College	High-school Graduate
Level I	15 −(21.1) − 6.1	22 −(25.1) − 3.1	30 −(20.9) 9.1
Level II	25 −(31.7) − 6.7	31 −(37.8) − 6.8	45 −(31.5) 13.5
Level III	30 −(29.2) 0.8	43 −(34.8) 8.2	20 −(29.0) − 9.0
Level IV	40 −(28.0) 12.0	35 −(33.3) 1.7	14 −(27.7) − 13.7

1. *Hypothesis to be tested.* The characteristics "sales volume" and "educational attainment" are independent
2. *Alternative hypothesis.* The characteristics are not independent

Decision rule If the chi-square statistic computed from the sample data is greater than 16.812 reject the hypothesis tested, accept the alternative hypothesis, and conclude that sales volume depends on educational attainment. Otherwise, conclude that sales volume does not depend on the agent's level of education.

The chi-square statistic computed from the sample data is 31.287, which exceeds the critical value. Therefore, we conclude that an agent's sales volume is related to his level of education.

When a *chi-square test of independence* leads to the conclusion that two variables (characteristics) are related, what does this tell us? Such a conclusion indicates that a relationship exists, but it provides no information about *how strong* this relationship is or *just how* these characteristics are related. For example, in our illustration, the test led us to the conclusion that sales volume and educational attainment are related. Do these characteristics have a strong relationship or only a weak relationship? Furthermore, does higher educational level mean higher sales volume or lower sales volume? The chi-square test of independence is a widely used and highly useful test. However, it does have these shortcomings. In chapters 19–22 we will study how to measure the *strength of relationship* and the *direction of relationship.*

Nevertheless, examination of the *pattern of differences* in Table 18.3 provides some insight into the direction as well as the strength of relationship between the two characteristics. Notice that for college graduates the differences are negative for sales levels I and II, indicating that (if the two characteristics were independent) fewer agents than expected are in these categories. On the other hand, the differences are positive for sales levels III and IV, indicating that more agents than expected are in the two higher sales levels. This indicates that, in general, college graduates produce higher levels of sales than would be expected if educational attainment is not actually related to sales level.

On the other hand, for high-school graduates, the reverse is true. The differences are positive for sales levels I and II and negative for sales levels III and IV. This indicates that, in general, high-school graduates do less well than expected in sales volume (assuming independence). The sample data indicate that higher educational attainment leads to higher sales volume. Furthermore, the consistency in the pattern of differences as well as their size indicates that there may be a strong relationship between the characteristics.

(**Note 18C** at the end of the chapter shows how to construct the chi-square test for a contingency table.)

18.3 Comparing a Distribution to a Standard

Testing the *goodness of fit* of a set of data to a *distribution standard* is another frequently used statistical test. For example, a market analyst claims that detergents A, B, and C have equal shares of the market and that each has a 30% share. In this illustration, the distribution of detergent sales as conceived by the analyst is A = 30%, B = 30%, C = 30%, other detergents = 10%. This can be tested by selecting a random sample of homemakers and comparing the distribution of the detergents they purchase with the foregoing distribution as the "standard for comparison" for the test.

The Business Conditions Analysis Group at a local university studied the income distribution in a large metropolitan area and constructed the distribution for the area as presented in Table 18.4. The long-range planning department of a national retail chain is studying the possibility of locating a branch store in this area. They are interested in testing the validity of the income distribution constructed by the Business Conditions Analysis Group. They select a random sample of 300 families in the metropolitan area and construct the income distribution presented in Table 18.5, column 2. Would you conclude, at the 0.05 significance level, that the income distribution in the metropolitan area is as constructed by the Business Conditions Analysis Group? The appropriate hypotheses for this test are as follows:

1. *Hypothesis to be tested.* The proportions of families in the five income categories are as determined by the Business Conditions Analysis Group
2. *Alternative hypothesis.* The proportions in the categories are not as determined by the Business Conditions Analysis Group

Notice in Table 18.5 that the *expected numbers of families* are presented in column 4. These expected frequencies are computed by multiplying the total number of families

Table 18.4

Income Distribution in Metropolitan Area as Constructed by Business Conditions Analysis Group

Annual Family Income	Proportion of Families in the Metropolitan Area
Under $1,000	0.10
$1,000–$2,499	0.20
2,500– 5,999	0.25
6,000– 9,999	0.20
10,000 and over	0.25
Total	1.00

Table 18.5

Income Distribution for a Sample of 300 Families, a Distribution Standard, the Expected Distribution, and Differences Between the Observed and Expected Distributions

Col 1	Col 2	Col 3	Col 4	Col 5
Annual Family Income	Sample Distribution (Number of Families)	Distribution Standard (Percent)*	Expected Distribution (Number of Families)†	Difference (Number of Families)‡
Under $1,000	22	10	30	−8
$1,000–$2,499	54	20	60	−6
2,500– 5,999	83	25	75	8
6,000– 9,999	74	20	60	14
10,000 and over	67	25	75	−8
Totals	300	100	300	

*From Table 18.4
†300 × Column 3
‡Column 2 − Column 4

in the sample (300) by the percentages in column 3 (which assumes that the column 3 distribution is valid). For example, 10% of 300 (or $0.10 \times 300) = 30$, etc. The observed (sample) number of families in each category is shown in column 2. The differences, *observed frequency − expected frequency*, are shown in column 5. The question to be answered is: Are these differences merely the result of random sampling or are they large enough to warrant the conclusion that the sample data distribution does not support the null hypothesis?

Decision rule If the chi-square statistic computed from the sample data is greater than 9.488, reject the hypothesis tested, accept the alternative hypothesis, and conclude that the income distribution in the area is not like the standard. Otherwise, conclude that the income distribution in the area is the same as the standard.

The chi-square statistic computed is 7.706, which is less than the critical value. We conclude that the distribution constructed by the Business Conditions Analysis Group may be taken to represent the income distribution in the metropolitan area.

(**Note 18D** at the end of the chapter shows how to construct a chi-square test of goodness of fit.)

18.4 Choosing Among Alternatives

The test procedure discussed in this section is the frequently used *analysis of variance.* This is a highly useful statistical test when it is desired to choose among alternatives. For example, will a new breakfast food sell better if it is packaged in small-, medium-, or large-size boxes? Is employee productivity higher if no coffee breaks, one, or two coffee breaks are permitted? Do students learn better if they are taught by method A, B, or C?

Analysis of variance is used to evaluate *experimental outcomes* where an activity is performed under a variety of conditions. Statisticians call the conditions imposed on the experiment *treatments.* Say a large facial tissue manufacturer has to decide whether to purchase machine A, B, or C for use in his production line. He could select three random samples of workers and assign each sample a set of identical tasks to be done. One sample would use machine A; another, machine B; and the third, machine C. The treatment (variable condition) in this experiment is the machine used. We may call the condition that is varied the *criterion of classification.*

The general approach of the analysis of variance test is to select the conditions (treatments) to be studied and to try out each condition on an independently selected sample. Then the mean outcome is computed for each sample and the sample means are compared to determine whether the variation in conditions results in differences in sample outcomes. In our example, the mean number of work units produced by each sample of workers is computed and the sample means are compared to evaluate whether the different machines have different effects on productivity.

Analysis of variance tests involve the *comparison of means.* Such tests vary from the simpler tests (which we will study) to highly complex *experimental designs.* The simplest tests involve only one criterion of classification and so are called *one-way analysis of variance* tests. More complex tests involve two or more classification variables. For example, if the workers in our example were separated by age category as well as machine used, we would have two variables of classification and the test procedure would be more complicated.

Suppose three methods of planting soybean seeds (A, B, C) are being studied by an agricultural scientist. Assume 15 plots equal in size and other characteristics are selected. Soybean seeds are planted in 5 plots using method A, 4 plots using method B, and 6 plots using method C. The number of bushels of soybeans harvested from each plot is shown in Table 18.6. Using the 0.05 level of significance, should it be concluded that one method is no better than another for planting soybean seed?

Table 18.6

Number of Bushels of Soybeans Harvested in Each of 15 Plots Planted by Methods A, B, and C

	Method A	Method B	Method C
	8	5	3
	7	4	2
	6	6	2
	4	2	1
	7		3
			4
Column Mean	6.40	4.25	2.50

Notice that in the chi-square tests discussed in the earlier sections of this chapter, we were concerned with *frequencies* (count data). As indicated in Table 18.6, in analysis of variance tests we deal with measurement *observations* such as "number of bushels harvested." As can be seen in Table 18.6, the observations vary from *column to column* and *within columns*. Analysis of variance tests are based on an analysis of the *total variation* in the observed data. More specifically, we are interested in partitioning the total variation of the data into meaningful and useful components. In one-way analysis of variance tests, total variation in the observed data is partitioned into two components, *variation between treatments* (or *variation between columns or samples*) and *variation within treatments* (*variation within columns or samples*).

The *total variation in the observed data* is determined by the variation of each observation (number of bushels harvested or the yield per plot) around the mean of all the observations. In our illustration, the mean yield of the 15 plots in the three samples combined is computed by dividing the sum of the 15 yields (64) by 15, obtaining 4.27 bushels per plot. In analysis of variance, this overall mean is called the *grand mean* of the observations.

We would expect the 15 soybean yields to vary around the grand mean because of the *different effects* of the three methods on yield as well as because of *chance effects* due to random selection of plots assigned to each planting method and the seeds used per plot. Of course, if the three methods have the *same effect* on yield, then method of planting would make no contribution to total variation. In such event, *total variation in the observed data would reflect only the variation due to chance.*

Notice in Table 18.6 that the mean yields per sample (6.40 bushels for method A, 4.25 bushels for method B, and 2.50 bushels for method C) differ from the grand mean yield (4.27 bushels). We would expect the sample (column) means to vary around the grand mean partly (perhaps mainly) because of the *different effects* of the planting methods on yield and partly because of *chance effects* related to random selection of plots and seeds. However, if planting methods were *equally effective* in soybean yield, then planting methods would make no contribution to the variation among sample means. In such an event, the *variation between treatments* (*samples*) *would reflect only chance variation.*

Notice in Table 18.6 how the observations in each column (sample) vary around their mean. We would expect the yields within each sample to vary around the sample mean *only* because of the *chance effects* of random sampling. Clearly, *no part* of this variation could be related to planting methods because the same method was used for all plots in a given sample. This is the variation within treatments. Statisticians also call this *error* (*or chance*) *variation.*

The statistician looks upon the three samples in Table 18.6 as representing three populations. The five yields shown for method A represent a sample of the population of yields that would be obtained if all soybean planting used method A (similarly for the sample yields for methods B and C).

The analysis of variance test is based on a comparison of the *variation between treatments* (*samples*) and *error* (*chance*) *variation.* If planting methods have *different effects* on soybean yield, then *between treatment variation should be larger than the error variation.* The important consideration is whether between treatment variation is sufficiently larger to warrant the conclusion that planting methods have different effects on yield and that the differences do not reflect merely chance differences. On the other hand, if the methods all have the *same effect* on yield, it would be expected that *between treatment variation and chance variation would be about the same.* The appropriate hypotheses are:

1. *Hypothesis to be tested.* The population mean yields (for methods A, B, and C) are all the same
2. *Alternative hypothesis.* The population mean yields are not all the same

Decision rule If the F statistic computed from the sample data is greater than 3.89 reject the hypothesis tested, accept the alternative hypothesis, and conclude that the three planting methods do not have the same effect on soybean yield. Otherwise, conclude that one method is just as good as another.

Based on the sample data, the F statistic is 10.64, which exceeds the critical value. Conclude that the planting methods have different effects on soybean yield.

(**Note 18E** at the end of the chapter shows how to construct a one-way analysis of variance test.)

Review

Concepts and Terminology

Alternatives
Analysis of variance
Cell
Joint-classification cell
Mean outcome
Measurement observations

Concepts and Terminology (*cont.*)

Chance effects
Chance variation
Chi-square statistic
Chi-square tests
Classification variables
Contingency table
Count data
Criterion of classification
Dependent
Different effects
Direction of relationship
Distribution standard
Error variation
Expected distribution
Expected frequencies
Expected number
Experimental design
F statistic
Goodness of fit
Grand mean
Independent
Joint-classification category
Observation data
Observations
Observed distribution
Observed frequencies
Observed number
One-way analysis of variance
Partitioning the total variation
Response categories
Same effects
Strength of relationship
Test of independence
Total variation
Treatments
Two-way classification frequency table
Variable conditions
Variation
Variation between columns
Variation between samples
Variation between treatments
Variation within columns
Variation within samples
Variation within treatments

Problems

1. Sample surveys on the outlook of business executives on conditions in the economy next year were conducted in four industries and the following summary of findings was prepared:

Number of Executives Who Responded

Response Category	*Forest Industry*	*Iron and Steel Industry*	*Textile Industry*	*Food Industry*
Improve	100	50	90	110
Not improve	600	450	910	690

(a) Identify the populations involved in these surveys.
(b) Determine the sample proportion of "improve" responses in each industry.
(c) If the proportion of "improve" responses is the same in the four populations included in the surveys, what would you estimate this proportion to be?
(d) Formulate the appropriate hypotheses for using these data to test whether or not the proportions of executives who expect economic conditions to improve is the same for the four industries.

2. In Problem 1, the following decision rule is constructed to test whether or not the four proportions are equal (at the 0.01 level of significance): If the chi-square statistic computed from the four samples exceeds 11.345 reject the hypothesis that the proportions are equal and accept the alternative hypothesis that they are not all equal. Otherwise, conclude that the proportions are equal.

(a) What is the critical value of the chi-square statistic in this test?

(b) Make the decision, if the chi-square statistic computed from the samples is 30.482.

3. What is meant by a test of independence? Give an illustration.

4. What is a contingency table?

5. A random sample of 300 adult consumers was selected in a large city. Each was asked his degree of preference for trading in a certain department store and also his degree of satisfaction in doing business in that store. The following table summarizes the replies:

Degree of Satisfaction	*Degree of preference*		
	High	*Medium*	*Low*
Not satisfied	10	20	40
Neutral	50	40	55
Satisfied	25	30	30

(a) What is the estimated probability that a consumer is neutral in degree of satisfaction?

(b) What is the estimated probability that a consumer has a high degree of preference for trading in the store?

(c) What are the variables of classification in this analysis?

(d) If the variables of classification are independent, what is the estimated probability that a consumer is both satisfied and has a low preference for trading in the department store?

(e) If the variables of classification are independent, what is the expected number of a sample of 300 consumers who would fall in the *satisfied-low preference* joint classification category?

(f) Formulate the appropriate hypotheses for conducting a test of independence.

6. In Problem 5, the following decision rule is formulated: If the chi-square statistic computed for the sample of 300 consumers exceeds 13.277, reject the hypothesis of independence and conclude that degree of preference and degree of satisfaction are dependent. Otherwise, conclude that these characteristics are independent. (The 0.01 significance level was used.) Apply the decision rule and reach a conclusion, if the chi-square statistic computed for these data equal 13.22.

7. The following table was taken from a trade journal:

Color of Luggage	*Distribution for a City*	*Distribution for the Country*
White	200	0.30
Blue	70	0.10
Green	60	0.15
Tan	120	0.30
Red	40	0.10
Others	10	0.05
	Total 500	Total 1.00

(a) Formulate the hypotheses appropriate for testing the goodness of fit of the distribution of luggage sold by color in the country to the distribution for the city, based on the random sample of 500 pieces of luggage sold in the city.

(b) Compute the expected numbers of luggage sold by color for the sample of 500, if the distribution for the city is the same as for the country.

(c) The following decision rule was formulated to test the goodness of fit of the distribution for the country to the distribution for the city (at the 0.05 level of significance): If the chi-square statistic computed from the sample data exceeds 11.070, reject the hypothesis that the city distribution is like the country distribution and conclude that the distributions are different. Otherwise, conclude that the distributions are the same. If chi-square is computed as 44.667 for the sample data, what conclusion would you reach?

8. What is a basic difference in the data used for chi-square tests and for the analysis of variance?

9. In analysis of variance tests, what is meant by criterion of classification? treatments?

10. What is a one-way analysis of variance test?

11. One-way analysis of variance tests are based on a partitioning of the total variation in the observed data into two components. Identify the two components.

12. What is the grand mean in an analysis of variance test?

13. In a one-way analysis of variance, explain the meaning of the following:
(a) total variation
(b) between treatment variation
(c) within treatment variation
(d) error variation

14. In a one-way analysis of variance test, if the treatments all have the same effect—
(a) what does the total variation in the observed data represent?
(b) what does the variation among the treatment means represent?

15. Explain why the variation within treatments cannot be explained by the variation in effects of the different treatments.

16. Analysis of variance is based on the comparison of between treatment variation and within treatment variation. Explain briefly.

17. A sample of four salesmen was randomly selected from each of the three branches of a firm and the number of orders written by each salesman during the past week was determined as follows:

Branch I	*Branch II*	*Branch III*
30	20	30
25	15	20
20	15	10
15	10	10

It is desired to use these sample data to determine whether the branches have equally effective sales forces.
(a) Compute the mean number of orders written per salesman in each branch.
(b) Compute the grand mean.
(c) What are two possible explanations for the differences between the grand mean and the sample means?
(d) What would explain the differences in the number of orders written by the salesman selected from a branch?

18. In Problem 17, formulate the appropriate hypotheses for a one-way analysis of variance.

19. In Problem 17, the decision rule is constructed as follows: If the F statistic computed from the samples exceeds 4.26 reject the hypothesis that the three branches have equally effective sales forces and accept the alternative hypothesis that the sales forces are not equally effective. Otherwise, conclude that the sales forces are equally effective. (The 0.05 level of significance was used.) Make the decision if the F statistic based on the sample data is 1.17.

Note 18A

Chi-square Tests and Computation of Chi-square Statistics

The χ^2 tests presented in chapter 18 compare *observed (sample) frequencies* (or counts) for a set of categories with corresponding *expected frequencies* determined on the basis of a hypothesis or on a theoretical basis. The χ^2 distribution was discussed in **Note 16C**.

The shape of the χ^2 distribution pertinent to a problem is determined by the number of categories involved. Actually, based on the number of categories, we compute the "degrees of freedom" (abbreviated as df) pertinent to the problem, and it is the number of df which determines the shape of the distribution. **Notes 18B, 18C,** and **18D** show how to determine the df for three types of problems.

The sample data used in a chi-square test are usually tabulated in a "row by column" arrangement, as illustrated in Table 18A.1. This table contains three rows and three columns and

Table 18A.1

Number of Families in Samples From Three Communities, by Income Level

Income Level	Community		
	Oceanview	Sunnyside	Lakeshore
Low	50	80	40
Medium	25	60	30
High	20	30	15

$3 \times 3 = 9$ *categories* or *cells.* Each cell contains a frequency or count based on the sample observations. A table such as Table 18A.1 is often called a 3×3 (3 by 3) table.

We will denote the observed frequency in a cell as X and the corresponding expected frequency as E. **Notes 18B**, **18C**, and **18D** show how to determine the expected frequencies for a test. The computation of the chi-square statistic is based on the difference between the observed and expected frequencies in the categories, denoted as $(X - E)$. More specifically, the chi-square statistic is computed as follows:

$$\chi^2 = \sum \frac{(X - E)^2}{E}$$

This equation indicates that for each cell we compute the difference $(X - E)$, square it to obtain $(X - E)^2$, and then divide by E to obtain the result $(X - E)^2/E$. This computation is made for each cell. Then, the symbol

$$\sum \frac{(X - E)^2}{E}$$

stands for "the sum of the results $(X - E)^2/E$ obtained for all the cells" in a problem. The computation of the chi-square statistic is illustrated in **Notes 18B**, **18C**, and **18D**.

Notice that the closer in value X and E are to each other, the smaller is the numerator $(X - E)$, and so the smaller is the χ^2 statistic. Consequently, the smaller the χ^2 statistic computed for a problem, the closer the agreement between the observed frequencies in the categories based on the sample data and the expected frequencies based on theoretical considerations or a hypothesis. In other words, the smaller the value of χ^2, the more confidence we have that the theory or hypothesis is true. The chi-square tests presented in chapter 18 use only upper-tail critical regions.

Note 18B

Chi-square Tests for Comparison of Population Proportions

(Before studying this note, it is necessary to study **Note 18A**.)

A random sample of adults was selected in each of three large metropolitan areas to compare the proportions of adults in these areas who feel that the U.S. Congress did not do a good job last session. The number in each sample who responded no to the question "Did the Congress do a good job?" and the number who responded yes are shown in Table 18B.1. Would you conclude, at the 0.005 significance level, that the proportion of adults who feel that the Congress did not do a good job is the same for the three metropolitan areas?

Table 18B.1

Congress Did a Good Job	Metropolitan Area			
	A	B	C	Total
No	50	160	80	290
Yes	300	340	370	1010
Totals	350	500	450	1300

In this problem, we have three populations (the adults in the three areas) and two categories for each population. Each column in Table 18B.1 relates to one sample and each row relates to a response category. Then, the table has 6 categories or cells (2 rows × 3 columns). The "no" category is the category of interest in this survey. The appropriate hypotheses are:

1. *Hypothesis to be tested.* The proportion of adults who feel that the Congress did not do a good job is the same in the three areas
2. *Alternative hypothesis.* The proportions are not all the same

Assuming that the null hypothesis is true, we estimate the common population proportion as follows:

$$\frac{\text{Total sample frequency in the category of interest for all samples combined}}{\text{Total number of observations in all samples combined}} = \frac{290}{1300} = 0.223$$

In other words, based on the frequencies in the "Total" column in Table 18B.1, 0.223 or 22.3% of the responses were no for the three samples combined.

Then, if the null hypothesis is true (and the population proportions are all the same), the expected number of no responses for the sample of 350 adults in area A is $0.223 \times 350 = 78.05$; for the sample of 500 adults in area B, it is $0.223 \times 500 = 111.50$; and for the sample of 450 adults in area C, it is $0.223 \times 450 = 100.35$. These are the *expected frequencies* (E) corresponding to the *observed frequencies* (X) shown in the "No" row of Table 18B.1. The expected frequencies for the "Yes" row are computed by subtraction (column total − expected "No" frequency). Table 18B.2 presents the observed and expected frequencies and computational details for computing the chi-square statistic.

Table 18B.2

No	A	B	C
X	50	160	80
E	78.05	111.50	100.35
$X - E$	−28.05	48.50	−20.35
$\frac{(X-E)^2}{E}$	10.0807	21.0964	4.1268
Yes			
X	300	340	370
E	271.95	388.50	349.65
$X - E$	28.05	−48.50	20.35
$\frac{(X-E)^2}{E}$	2.8932	6.0547	1.1844

Finally, the χ^2 statistic is computed as follows:

$$\chi^2 = 10.0807 + 21.0964 + 4.1268 + 2.8932 + 6.0547 + 1.1844$$
$$= 45.4362$$

Since the level of significance for this problem is 0.005, refer to Table B.2 to find a χ^2 value which cuts off 0.005 of the area under the χ^2 distribution in the upper tail. As discussed in **Note 18A**, the greater the differences between X and E in the categories related to a problem, the larger is χ^2 and the less confidence we have that the null hypothesis is true. Therefore, in tests of the equality of proportions, the critical region is always in the upper tail.

The column headed "0.005" in Table B.2 presents the critical χ^2 values for this problem. How do we determine the number of degrees of freedom? In tests comparing several proportions, df =

number of samples − 1. For our problem, df = 3 − 1 = 2. Therefore, referring to Table B.2, column headed "0.005," for 2 df, $\chi^2 = 10.597$.

Decision rule If the chi-square statistic computed from the sample data is greater than 10.597, reject the hypothesis tested, accept the alternative hypothesis, and conclude that the population proportion is not the same for the three metropolitan areas. Otherwise, conclude that they are the same.

The chi-square statistic computed from the samples (45.4362) clearly exceeds the critical value. Therefore, we conclude that the proportions in the three areas are not the same.

Note 18C

Chi-square Tests for a Contingency Table

(Before studying this note, it is necessary to study **Note 18A**.)

A random sample of 300 podiatrists provided information on "number of years active in the profession" and "professional activity." Table 18C.1 presents the number of sample observations in each of the 9 cross-classification cells of the contingency table constructed for these data. This is a 3 × 3 contingency table. Based on these data, would you conclude at the 0.01 level of significance that "number of years active" and "professional activity" are independent for podiatrists? The appropriate hypotheses are:

1. *Hypothesis to be tested:* Professional activity and years active in podiatry are independent
2. *Alternative hypothesis:* These characteristics are not independent

Table 18C.1

Professional Activity	Years Active in Podiatry			
	0–9	10–19	20 and over	Total
Teaching	20	40	10	70
Patient care	40	55	50	145
Other activity	30	30	25	85
Totals	90	125	85	300

Consider the cross-classification cell in Table 18C.1 for "teaching and 0–9 years active" which shows an observed frequency (X) of 20 podiatrists. Assuming that the hypothesis of independence (the null hypothesis) is true, according to probability Rule D for independent events (**Note 10A**) we may write the following:

$$P(\text{teaching and 0–9 years active}) = P(\text{teaching}) \times P(\text{0–9 years active})$$

The terms of this probability equation may be interpreted as follows:

1. P(teaching) denotes the probability that if a podiatrist is selected at random, he will be engaged in teaching. Using the sample data in the total column of Table 18C.1, we estimate this probability as follows:

$$P(\text{teaching}) = \frac{\text{Total of the "teaching" row}}{\text{Sample size}} = \frac{70}{300}$$

2. P(0–9 years active) denotes the probability that if a podiatrist is selected at random, he will have been active in the profession for 0–9 years. Using the sample data in the total row of Table 18C.1, we estimate this probability as follows:

$$P(\text{0–9 years active}) = \frac{\text{Total of the 0–9 column}}{\text{Sample size}} = \frac{90}{300}$$

3. P(teaching and 0–9 years active) denotes the probability that if a podiatrist is selected at random, he will be both engaged in teaching and have 0–9 years active in the profession. Assuming these represent independent events (according to the null hypothesis) we estimate this probability by multiplying the foregoing results as follows:

$$P(\text{teaching and 0–9 years active}) = \frac{70}{300} \times \frac{90}{300} = 0.07$$

Recall that a probability has a relative frequency interpretation (chapter 10); so the result just obtained indicates that 0.07 (or 7%) of the podiatrists are estimated to fall in the cross-classification category "teaching and 0–9 years active." Therefore, the expected frequency for this category is computed as 0.07 × sample size ($n = 300$). Or,

$$\frac{70}{300} \times \frac{90}{300} \times 300 = \frac{70 \times 90}{300} = 21.0$$

In general, we compute the expected frequency (E) corresponding to the observed frequency (X) for a category as follows:

$$E = \frac{\begin{pmatrix}\text{Corresponding}\\ \text{row total}\end{pmatrix} \times \begin{pmatrix}\text{Corresponding}\\ \text{column total}\end{pmatrix}}{\text{Sample size}}$$

Applying the equation to the data in Table 18C.1, we obtain for the cell in the first row, first column (where $X = 20$), the following:

$$E = \frac{70 \times 90}{300} = 21.0 \text{ (as before)}$$

For the cell in the first row, third column (where $X = 10$)

$$E = \frac{70 \times 85}{300} = 19.8$$

The other E values are computed similarly. The chi-square statistic for this problem is conveniently computed as shown in Table 18C.2.

Table 18C.2

X	E	$\frac{(X - E)^2}{E}$
20	21.0	0.05
40	43.5	0.28
30	25.5	0.79
40	29.2	3.99
55	60.4	0.48
30	35.4	0.82
10	19.8	4.85
50	41.1	1.93
25	24.1	0.03
	Total	$\chi^2 = 13.22$

As discussed in **Note 18A**, the greater the differences between X and E in the categories related to a problem, the larger is χ^2 and the less confident we are that the null hypothesis is true. Therefore, in a test of independence, the larger χ^2, the more confident we are that the hypothesis of independence is not true; so the critical region is always in the upper tail. The number of degrees of freedom in such a test is computed as follows:

$$\text{df} = (\text{number of rows} - 1) \times (\text{number of columns} - 1)$$

In our problem, df $= (3 - 1) \times (3 - 1) = 4$. Since the level of significance is 0.01, refer to Table B.2 (column headed "0.01") and read out, for df $= 4$, $\chi^2 = 13.277$, the critical value for the test.

Decision rule If the chi-square statistic computed from the sample data is greater than 13.277, reject the hypothesis tested, accept the alternative hypothesis, and conclude that "number of years active" and "professional activity" are related. Otherwise, conclude that these characteristics are independent.

As computed in Table 18C.2, $\chi^2 = 13.22$ falls in the region of acceptance. Therefore, we accept the hypothesis being tested and conclude that the two characteristics are independent. In other words, the professional activity of a podiatrist does not depend on the number of years he was active in his profession.

Note 18D

Chi-square Test of Goodness of Fit

(Before studying this note, it is necessary to study **Note 18A**.)

Tests relating to "goodness of fit" are based on a comparison of the distribution of a random sample of observations in a set of categories with a hypothesized distribution. For example, referring to Table 18.5 (Section 18.3), we find the sample of 300 families is distributed into a set of five income categories as shown in column 2 and the hypothesized distribution is shown in column 3.

The formulation of the hypothesis to be tested and the alternative hypothesis is shown in Section 18.3. The observed frequencies (X) are in column 2 (Table 18.5) and the expected frequencies (E) in column 4. The computation of the expected frequencies assumes that the null hypothesis is true.

Notice that (sum of the expected frequencies) = (sum of the observed frequencies) = 300 (the sample size). This is always true in a test of goodness of fit and represents a restriction imposed on the computation of the expected frequencies. The chi-square statistic is conveniently computed as shown in Table 18D.1.

As discussed in **Note 18A**, the larger the ($X - E$) differences, the larger is the chi-square statistic computed from the sample data and the less confidence we have that the hypothesis tested is true. Therefore, in a test of goodness of fit, the larger χ^2, the less confidence we have that the sample observations come from a population with a distribution as hypothesized. Hence, the critical region is always in the upper tail of the chi-square distribution in such a test. The number of degrees of freedom is computed as

$$\text{df} = (\text{number of categories}) - (\text{number of restrictions imposed})$$

Notice from Table 18.5 (or Table 18D.1) that the number of categories is 5 for our problem. The "number of restrictions imposed" relates to the computation of the expected frequencies.

Table 18D.1

X	E	$\frac{(X-E)^2}{E}$
22	30	2.133
54	60	0.600
83	75	0.853
74	60	3.267
67	75	0.853
	Total	$\chi^2 = 7.706$

As previously explained, only one restriction was imposed in our problem (sum of expected frequencies = sample size). Therefore, for our problem, df = 5 − 1 = 4. Often, it is desired to test whether a population distribution is similar to a theoretical distribution, such as the normal distribution. In such problems, the number of restrictions imposed on the computation of the expected frequencies is greater than one. However, we will limit our study of tests of goodness of fit to problems which involve only one restriction.

Since the level of significance is 0.05 (as specified in Section 18.3) and df = 4, refer to Table B.2, column headed "0.05" and read out $\chi^2 = 9.488$ as the critical value for our test. Then, the decision rule is expressed as in Section 18.3 and the decision is made as shown.

Note 18E

One-way Analysis of Variance Tests

(Before studying this note, it is necessary to study **Note 17D** at the end of chapter 17.)

A firm produces three types of chemical plant sprays (I, II, III). Three random samples of a certain variety of ornamental plant were selected and spray I was used on the first sample, spray II on the second sample, and spray III on the third sample. Each sample of plants contained four flowers. During a certain period of time, the number of bees which lighted on each of the flowers in each sample was counted, as shown in Table 18E.1. Would you conclude, at the 0.05 level of significance, that the three sprays are equally effective in attracting bees to the flowers?

Table 18E.1

	Chemical Plant Spray Used			
	I	II	III	
	6	10	7	
	5	11	9	
	10	9	10	
	7	9	9	
T	28	39	35	$\Sigma\, T = 102$
n	4	4	4	$\Sigma\, n = 12$
$\bar{X}$	7.00	9.75	8.75	$G\bar{X} = 8.5$
$\frac{T^2}{n}$	196.00	380.25	306.25	$\sum \frac{T^2}{n} = 882.50$
TSQ	210	383	311	$\Sigma\, TSQ = 904$

Notice from Table 18E.1 that for the sample of flowers sprayed with spray I, one flower attracted six bees, another five bees, etc. This table also includes some of the computations involved in conducting the statistical test. The meaning of the symbols shown in the table are:

T = a column (sample) total
ΣT = sum of the T's
n = sample size
Σn = sum of the n's
$\bar{X}$ = sample mean
$G\bar{X}$ = grand mean (computed as $\Sigma T/\Sigma n$).
$\dfrac{T^2}{n}$ = the square of T for a sample divided by n
$\sum \dfrac{T^2}{n}$ = sum of the $\left(\dfrac{T^2}{n}\right)$'s
TSQ = sum of the squares of the observations in a column (sample)
For the spray I sample, $TSQ = 6^2 + 5^2 + 10^2 + 7^2 = 210$
$\Sigma\ TSQ$ = sum of the TSQ's

In conducting an analysis of variance test, it is assumed that the samples were independently selected from normally distributed populations with equal population variances (σ^2). Hypotheses are formulated relating to the population means (μ's) as follows:

1. *Hypothesis to be tested.* $\mu_1 = \mu_2 = \mu_3$ (the three population means are all equal)
2. *Alternative hypothesis.* The μ's are not all equal

The null hypothesis implies that the different treatments (sprays) all have the same effect, since all sprays result in the same mean number of bees attracted. Alternatively, if the sprays have different effects on the number of bees attracted, it is to be expected that the population means would not all be equal (as stated in the alternative hypothesis).

If the null hypothesis is true, normal populations with equal treatment means and a common variance σ^2 may be considered to make up a single normal population with mean μ and variance σ^2. The test procedure is based on two estimates of σ^2 computed from the sample data on the assumption that the null hypothesis is true, as follows:

$$MSE = \frac{SSE}{\text{df for } SSE}$$

and

$$MST = \frac{SST}{\text{df for } SST}$$

Let us consider the meaning of these new symbols and equations. *SSE* is an overall measure of the variation of the individual observations (the number of bees lighting on each flower) from their sample mean. For example, for the sample tested with spray I, this measure takes into account the variation of 6, 5, 10, and 7 bees from the mean of 7.00 bees; for the sample treated with spray II, this measure takes into account the variation of 10, 11, 9, and 9 bees from the mean of 9.75 bees; etc. The df associated with *SSE* is computed as (Σn − the number of treatments). In our problem df for $SSE = 12 - 3 = 9$.

SEE is called technically the *sum of squares within treatment samples* because it is computed based on the square of the deviation of each observation from its treatment mean. For example, for the first observation for spray I, we have 6 bees and the sample mean is 7.00 bees. Therefore, the deviation is $6 - 7 = -1$ and the square is $-1 \times -1 = 1$. We would expect that the observations in a given sample would vary around the sample mean only because of chance. Clearly, since the same spray was used for all the flowers in a given sample, no part of this variation reflects different effects of the sprays. Consequently, this variation represents experimental error (chance) variation, so *SSE* stands for *sum of squares for error*.

In analysis of variance tests, a variance is called a *mean square*. When *SSE* is divided by its associated degrees of freedom, we call the result the *mean square for error* and denote this as *MSE*.

SST stands for the *sum of squares for treatments* and represents a measure of the variation of treatment sample means around the grand mean $G\bar{X}$. This sum of squares is computed based on the square of the deviation of each sample mean from the grand mean. For example, for the spray I treatment sample, the mean is 7.00 bees and $G\bar{X} = 8.5$ bees. Therefore, the deviation is computed as $7 - 8.5 = -1.5$ and the square is $-1.5 \times -1.5 = 2.25$. The df associated with *SST* is computed as (number of treatments -1). In our problem, df $= 3 - 1 = 2$.

We would expect treatment sample means to vary around the grand mean mostly because of the different effects of the sprays, if the alternative hypothesis is true. If the null hypothesis is true, then *SST* would reflect only chance variation. When *SST* is divided by its associated df, we call the result the *mean square for treatments* and denote this as *MST*.

Applying the computational results from Table 18E.1, we compute the following:

$$SST = \sum \frac{T^2}{n} - \frac{(\Sigma\, T)^2}{\Sigma\, n}$$

and

$$SSE = \sum TSQ - \sum \frac{T^2}{n}$$

Using these equations, we compute *SST* and *SSE* for our problem as follows:

$$SST = 882.50 - \frac{(102)^2}{12} = 15.5$$

$$SSE = 904 - 882.50 = 21.5$$

Then, applying the previous equations, we compute the following:

$$MSE = \frac{21.5}{9} = 2.39$$

$$MST = \frac{15.5}{2} = 7.75$$

Finally, the F statistic is computed as follows:

$$F = \frac{MST}{MSE}$$

Applying this equation to our problem, we compute the F statistic as $7.75/2.39 = 3.243$.

How are these results interpreted and how is the decision rule constructed? If the null hypothesis is true, so that the treatment means are all equal (and the sprays have the same effect on attracting bees), *MSE* and *MST* both represent estimates of σ^2 (the variance due to chance). Then, they would be expected to be about equal. On the other hand, if the treatment effects are not the same (so that the alternative hypothesis is true), we would expect MST (which reflects the differences among treatment effects and chance) to be larger than MSE (which reflects chance effects only). Since $MST = 7.75$ is larger than $MSE = 2.39$, we need to determine whether MST is sufficiently larger to warrant rejection of the null hypothesis.

Since the level of significance is specified as 0.05, refer to Table B.3 in Appendix B and read out the critical F value of 4.26 for 2 df (numerator variance) and 9 df (denominator variance). Then, the decision rule is expressed as follows:

Decision rule If the F statistic computed from the sample data is larger than 4.26, reject the hypothesis tested, accept the alternative hypothesis, and conclude that the sprays have different effects on attracting bees. Otherwise, conclude that the sprays all have the same effect.

Noting that the computed F statistic is 3.243, which is less than 4.26, we conclude that the three sprays have the same effect.

Part E

ASSOCIATION AND PREDICTION

19

How the Statistician Looks Upon Relationship

The study of *relationship* is a most important application of statistical analysis. For example, a psychologist may wish to determine whether there is a relationship between stress and learning speed and whether the learning speed of a student can be predicted based on a measure of stress. Or, a business analyst may need to predict the demand for a product and attempt to base his prediction on a measure of the relationship between demand and average family income. Or, a sociologist may wish to investigate the relationship between the crime rate and the proportion of employed mothers of school-age children and whether this proportion provides a basis for predicting the crime rate in a community.

19.1 Regression and Correlation Analyses

Statisticians distinguish between two types of analyses in the study and measurement of relationship—*regression analyses* and *correlation analyses*. Consider the problem of the chemist who wishes to determine the relationship between reaction time of a chemical process and the temperature under which it takes place and also wishes to predict reaction time based on temperature. A statistical determination of relationship for this problem is based on data for the two variables "reaction time" and "temperature." Since it is desired to predict reaction time based on temperature, reaction time is the *predicted variable* and temperature is the *predictor variable* or, merely, the *predictor*.

Suppose the chemist decides to perform experiments under 10 *specified* temperatures and to note the reaction time in each case. This will provide 10 temperature values and 10 reaction times for statistical analysis. Notice carefully that the values of the predictor variable "temperature" are *chosen* by the chemist. These are *not* random values of this variable. On the other hand, the values of the predicted variable "reaction time" are obtained as a consequence of the experiment. Clearly, even for a given temperature, it is to be expected that reaction time would vary from experiment to experiment. Therefore, the values for the predicted variable *are* random values.

A statistical analysis of relationship where the predictor values (say, temperature) are not random values and the predicted variable values (say, reaction time) are random values, is called a *regression analysis*. The major purpose of a regression analysis is *prediction*. In such analyses, the statistician computes a *prediction equation*, called, technically, a *regression equation*. Statistical theory used to develop such an equation is said to be based on a *regression model of relationship*.

On the other hand, suppose an educator is interested in determining whether there is a relationship between a student's high-school average and his first-year college average. Here, the statistical analysis is also based on two variables, "high-school average" and "first-year college average." Say the educator selects a *random sample* of 20 college students and obtains for each student in the sample his high-school average and his first-year college average. The sample data provide him with 20 *random values* for *each variable*.

A statistical analysis of relationship where the values for *both* variables are random values is called a *correlation analysis*. An important objective of a correlation analysis is to *measure the nature and extent of association between the variables in the analysis*. In such analyses, the statistician usually computes a *coefficient of correlation* which provides a measure of *how closely* the variables are associated. For technical reasons this measure is called a *correlation index* in some analyses.

Very often, an additional objective of a correlation analysis is to determine a prediction equation as well as a measure of association. For example, in our illustration, the educator may wish to predict first-year college averages based on high-school averages as well as to determine how closely these variables are associated.

It is worth noting that in a correlation analysis in which the only objective is to measure how closely the variables are associated, statisticians do not refer to the variables as predictor and predicted variables. They are merely referred to as "variables." However, whenever prediction is an objective of an analysis, one variable is the predictor (the basis of the prediction) and the other is the predicted variable. Statisticians often call the predictor the *independent variable* and the predicted variable the *dependent variable*. Statistical theory used to develop a measure of association and, perhaps, also a prediction equation is said to be based on a *correlation model of relationship*

Suppose an economist wishes to study the relationship between the number of one-family homes sold in a year and the number of marriages performed three years earlier. He also wishes to determine a prediction equation to predict the number of home sales for a year based on the number of marriages. Say he *chooses* 50 particular communities each of about 10,000 population where the number of marriages performed is different for each community and ranges from some very low number to some very high number. Does this problem fit the regression model or the correlation model? Clearly, the 50 values for the predictor ("number of marriages performed three years earlier") were chosen by the economist and so are not random values. The 50 values for the variable "number of one-family homes sold in a year" are random values. Then, this is a regression analysis of relationship. However, if the economist selects a *random sample* of 50 communities from the population of all communities of about 10,000 population, then the values for both variables would be random values and we would have a correlation analysis.

19.2 Usefulness of Statistical Methods

In a *regression analysis* it is statistically valid to compute a prediction equation and measure how good this equation is as a basis for prediction. It is also possible to compute a measure of the degree of association between the variables in the analysis (the correlation coefficient). However, it is *not possible* to use statistical theory to evaluate how good this measure is. This can be done only on a judgmental basis. However, in a *correlation analysis* it is valid to apply statistical theory to evaluate the measure of association, as well as to evaluate how good a prediction equation has been obtained.

We will consider prediction equations obtained in a statistical analysis of relationship in chapter 20 and measures of association in chapter 21.

Review

Concepts and Terminology

Association
Coefficient of correlation
Correlation analysis
Correlation index
Correlation model
Dependent variable
Independent variable
Predicted variable
Prediction
Prediction equation
Predictor variable
Random values
Regression analysis
Regression equation
Regression model
Relationship

Problems

1. A business analyst carefully chooses a sample of manufacturing firms so that annual gross sales per firm is evenly distributed from low sales volume to high sales volume. He determines for each firm the average weekly wage paid to mechanics. The analyst wishes to determine whether average weekly wage paid to mechanics by a manufacturing firm can be predicted based on the firm's annual gross sales volume.

(a) Identify the two variables in this analysis and indicate which is the predictor and which is the predicted variable.
(b) Which variable is independent and which is dependent?
(c) Are both variables random variables or only one? If one, which is the random variable? Defend your answer.
(d) Is this a regression analysis or a correlation analysis? Why?

2. In Problem 1, suppose the business analyst selects a random sample of manufacturing firms and for each firm he determines annual gross sales and average weekly wage paid to mechanics. Assume the analyst has the same objective in his analysis as previously.
(a) Which is the dependent variable and which the independent variable?
(b) Are both variables random variables or only one? If only one, which is the random variable?
(c) Is this a regression analysis or a correlation analysis? Why?

3. A botanist wants to determine if he can predict the height of a certain potted plant after a specified period of growth based on the number of times the plant was watered during the specified period. He selects 20 plants with a wide range of watering times and determines the height of each plant.
(a) Identify the predicted and predictor variables in this analysis.
(b) Does this analysis fit the regression model or the correlation model? Why?

4. In Problem 3, suppose the botonist selected the 20 plants randomly from a large supply of such plants. Does the analysis fit the regression or the correlation model? Why?

5. A lawyer wants to determine whether his fee for a case is related to the number of days he must spend in court on the case. Suppose he selects a random sample of cases from his records of the past five years and determines, for each case selected, the fee charged and number of days spent in court. He plans to conduct an analysis based on these data. Does this analysis fit the regression model or the correlation model? Why?

6. In Problem 5, suppose the lawyer selects the sample of cases to be sure he gets a sample with a wide range of days spent in court. Does the analysis fit the regression model or the correlation model of analysis? Why?

7. A hospital administrator wants to determine whether there is a relationship between the number of days a patient spends in the hospital and the age of the patient. He also would like to be able to predict the number of days to be spent in the hospital based on age. He decides to study a sample of patients from the hospital records of the last ten years. How should he select the sample if he wants to conduct a correlation analysis?

8. What is a coefficient of correlation?

9. What is a more technical name for a prediction equation?

10. Suppose you want to compute the coefficient of correlation to measure the degree of association between two variables. You also want to evaluate how good a measure it is. Which model, the regression model or the correlation model, is appropriate? Why?

20

Prediction

20.1 Prediction Equations

Table 20.1 presents data on typing speed (words per minute) and typing experience (number of months of employment) for a sample of 12 typists. Suppose it is desired, using these data, to determine a prediction equation to predict typing speed based on experience. Then, the predictor is "experience" and the predicted variable is "typing speed". Figure 20.1 presents a *scatter diagram* for these data. The scatter diagram is a convenient and highly effective device used to display the relationship between two variables.

In a *scatter diagram*, a scale of values for the predictor is shown on the horizontal scale and a scale for the predicted variable is shown on the vertical scale. A dot is plotted for each *pair of values*. For example, the first pair of values in Table 20.1 is 33 (typing speed) and 4 (experience). Locating 4 on the horizontal scale, we plot a dot directly above at a height equal to 33 on the vertical scale. This is identified as dot A in Figure 20.1. The scatter diagram contains 12 dots, one for each pair of values in the table.

A scatter diagram is sometimes called a *dot chart*. Notice that the scatter of dots in Figure 20.1 suggests a *straight line*. Such a line has been drawn through the scatter of dots and is presented in Figure 20.2. This line represents the relationship between the two variables. It is called a *regression line* and represents the joint variation of the two variables—typing speed and experience. The regression line, which was constructed on a freehand (judgement) basis, may be used to predict typing speed based on the number of months of experience.

Table 20.1

Typing Speed (Words per Minute) and Experience (Number of Months of Employment) for a Sample of 12 Typists

Typing Speed	Experience
33	4
45	7
46	8
20	1
40	6
30	3
38	5
22	2
52	9
44	6
42	7
55	10

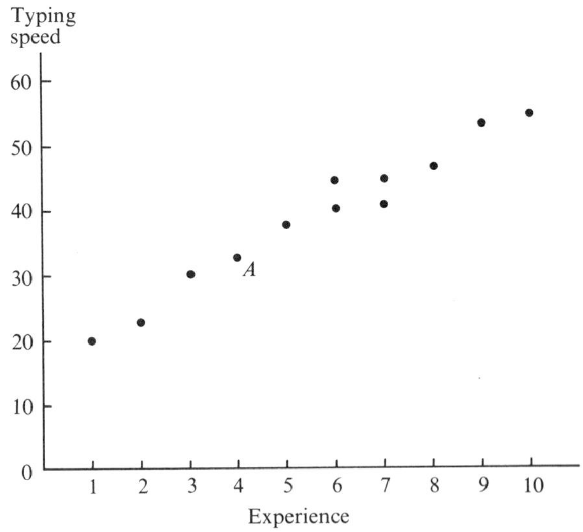

Figure 20.1

Scatter Diagram for the Data in Table 20.1

Assume it is desired to predict typing speed for a typist with 5 months of experience. Referring to Figure 20.2, we find 5 on the experience scale (horizontal). Then we move upward from this point (as shown by the dashed vertical line in the figure) until the regression line is reached. Now, we move to the left as indicated by the horizontal dashed line and read off about 36 words per minute as the predicted typing speed.

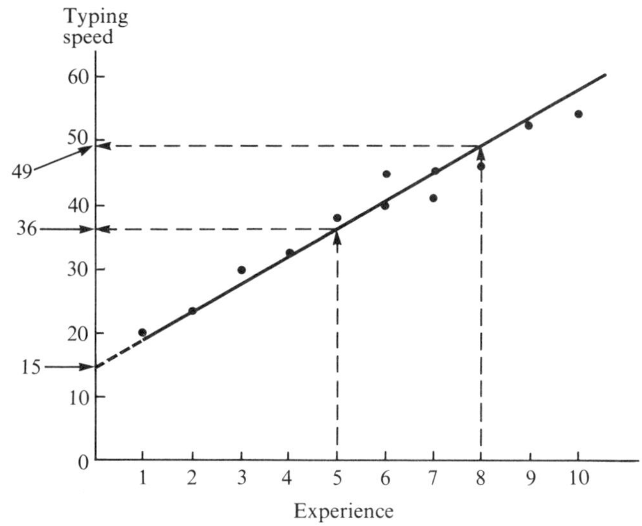

Figure 20.2

Scatter Diagram and Regression Line for the Data in Table 20.1

In other words, the relationship between typing speed and experience for the sample of 12 typists as represented by the free-hand regression line indicates that a typist with 5 months of experience may be expected to have a typing speed of 36 words per minute, on the average. In the same way, the predicted typing speed for a typist with 8 months of experience is about 49 words per minute as indicated in the figure. Summarizing, we have the following:

Months of Experience	*Predicted Typing Speed*
5	36
8	49

Taking differences, we obtain $8 - 5 = 3$, $49 - 36 = 13$ and $13/3 = 4.3$, which statisticians call the *slope* of the regression line. The *slope* indicates the increase in typing speed (4.3 words per minute) to be expected on the average for each additional month of typing experience.

We extend the regression line until it meets the typing speed (vertical) scale of values, as shown by the dashed line extension in Figure 20.2. At this point, experience = 0 months. As indicated in the figure, this extension meets the vertical scale at about 15 words per minute. Statisticians call this the *height* of the regression line. Putting the foregoing results together, we may write

$$\text{Predicted typing speed} = \text{height of the regression line} + \text{slope of the regression line} \times \text{number of months of experience}$$

Since height = 15 and slope = 4.3, we may write more elegantly

$$\text{Predicted typing speed} = 15 + 4.3 \times \text{number of months of experience}$$

We have just constructed a *prediction equation* for typing speed based on the data for the sample of 12 typists and the freehand regression line. Using this equation, it is quite simple to predict typing speed for any specified amount of experience. What would you predict as the typing speed for a typist with 9 months of experience, using the prediction equation? Applying the equation, we obtain the following:

$$\begin{aligned}\text{Predicted typing speed} &= 15 + 4.3 \times 9 = 15 + 38.7 \\ &= 53.7 \text{ words per minute}\end{aligned}$$

Thus, for a typist with 5 months of experience the predicted typing speed is 36.5 words per minute compared with 36 read off the figure. For 8 months, it is 49.4 words per minute compared with 49 read off the figure. The foregoing procedure for determining a regression line, the height and slope of this line, and a prediction equation is dependent on the judgement of the analyst and may be expected to vary from analyst to analyst. Such a line is satisfactory only when a rough prediction equation is desired. Generally, a more precise, objectively determinated prediction equation is required.

20.2 ''Best-fitting'' Regression Lines

Notice that the roughly determined prediction equation predicted a typing speed of 36.5 words per minute for a typist with 5 months of experience compared with an observed typing speed of 38 words per minute for a typist in the sample with 5 months of experience (Table 20.1). The difference (38 − 36.5 = 1.5 words per minute) represents the *deviation* between the *observed* and *predicted* typing speed. The predicted speed for a typist with 8 months of experience is 49.4 words per minute compared with an observed speed of 46 words per minute for such a typist in the sample, a deviation of 46 − 49.4 = −3.4 words per minute (observed speed is 3.4 words per minute *less than* the predicted speed). The predicted typing speed for a typist with 9 months of experience is 53.9 words per minute compared with an observed speed of 52 words per minute shown in Table 20.1, a deviation of 52 − 53.9 = −1.9 words per minute.

Statisticians generally determine a "best-fitting" regression line and compute a more precise prediction equation by *minimizing these deviations.* More specifically, the statistician looks at the *squares of the deviations* (which are computed as 1.5 × 1.5 = 2.25 for the deviation of 1.5 words per minute, −3.4 × −3.4 = 11.56 for the −3.4 words per minute deviation, etc). He determines the squares of the deviations for all the data in the scatter diagram and computes a regression line and a prediction equation which *minimizes the sum of the squares of the deviations.* These are called a *least squares regression line* and a *least squares prediction equation.*

Some deviations are *positive* (such as 1.5 words per minute) which indicates that the observed typing speed is *greater than* the predicted speed. Other deviations are *negative* (such as −3.4 words per minute) which indicates that the observed typing speed is *less than* the predicted speed. When a least squares regression line is determined, the positive and negative deviations just balance out. In other words, the sum of the positive deviations equals the sum of the negative deviations. (**Note 20A** at the end of the chapter shows how to compute a least squares prediction equation.)

Using the *method of least squares*, the following equation is obtained (as shown in **Note 20A**):

$$\text{Predicted typing speed} = 16.9 + 3.9 \times \text{number of months of experience}$$

Note that the slope (3.9) and height (16.9) differ from the previously obtained results. Nevertheless, the predicted typing speeds are not *too* different, as shown below:

	Predicted Typing Speed	
Experience	*Rough Prediction Equation*	*Least Squares Prediction Equation*
5	36.5	36.4
8	49.4	48.1
9	53.7	52.0

A prediction equation based on a freehand regression line often shows greater differences. Statisticians call the height and slope of a regression equation the *regression coefficients* and usually refer to the prediction equation as a *regression equation.*

20.3 Types of Relationships

Observe the scatter of dots and the regression line in Figure 20.2. Notice that low values of the predictor (experience) are associated with low values of the predicted variable (typing speed). In other words, when the experience variable has a low value, the typing speed variable has a low value. When two variables are so related, they are said to be *directly* or *positively related.* In such cases, the regression line slopes upward (from the lower left corner to the upper right corner) and is said to be *positively sloped.*

Table 20.2 presents the quantity (dozens) of a food product sold and the average selling price (cents per dozen) for the most recent 10 weekly periods. Suppose it is

Table 20.2

Quantity Sold (dozens) and Average Price (cents per dozen) for a Food Product for the Most Recent 10 Weekly Periods

Quantity Sold	Price
30	28
47	22
38	29
28	32
49	20
23	35
47	21
46	20
39	24
42	27

desired to predict quantity sold based on average selling price. Figure 20.3 shows the scatter diagram for these data. Notice that in this figure the scatter of dots indicate that low values of the predictor (price) are associated with high values of the predicted variable (quantity sold) and that the straight line suggested by the scatter of dots slopes downward from the upper-left corner to the lower-right corner.

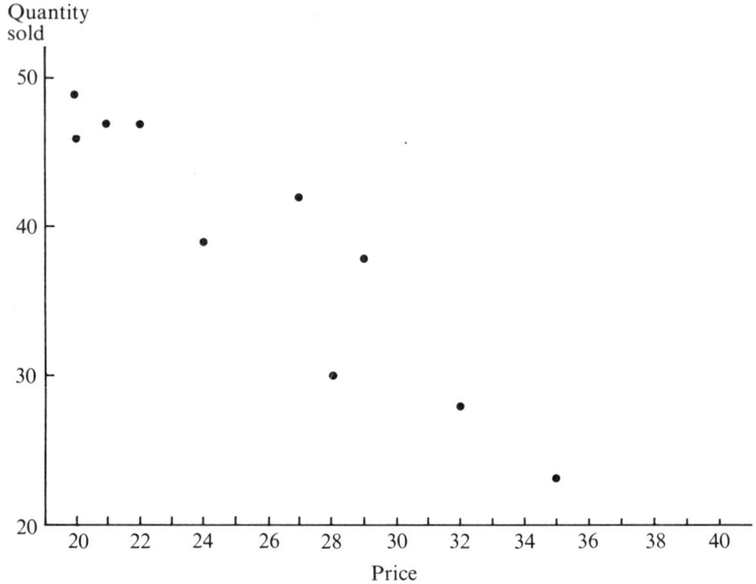

Figure 20.3

Scatter Diagram for the Data in Table 20.2

Figure 20.4 presents a freehand regression line drawn to fit the scatter of dots. When two variables are related as in Figure 20.4, they are said to be *inversely* or *negatively related* and the regression line is said to be *negatively sloped.*

Based on the freehand regression line, we may predict (as shown in Figure 20.4) that 38 dozen will be sold if the average price is 26¢ per dozen and 29 dozen will be sold if the average price is 31¢ per dozen. Then, taking differences, we obtain $31 - 26 = 5$¢ per dozen, $38 - 29 = 9$ dozen. The slope of the regression line is $9/5 = 1.8$. However, since the regression line is negatively sloped, we must write the slope as -1.8, which indicates that quantity sold is *reduced* 1.8 dozens for each one-cent *increase* in the average price. Or, conversely, the quantity sold is *increased* 1.8 dozens for each one-cent *reduction* in the average price.

Recall how the height of the freehand regression line was determined for Figure 20.2 where the regression line was extended to meet the vertical scale at the point where the predictor (experience) was equal to 0. This cannot be done so easily in Figure 20.4 because the point for 0 price does not appear on the chart. An alternative way to determine the height of the regression line is to note that at the average price of 26¢ per dozen we predicted 38 dozens will be sold. Then, multiplying 26×1.8 (the slope) we obtain 46.8 and $38 + 46.8 = 84.8$ dozen, which represents the height of the free-

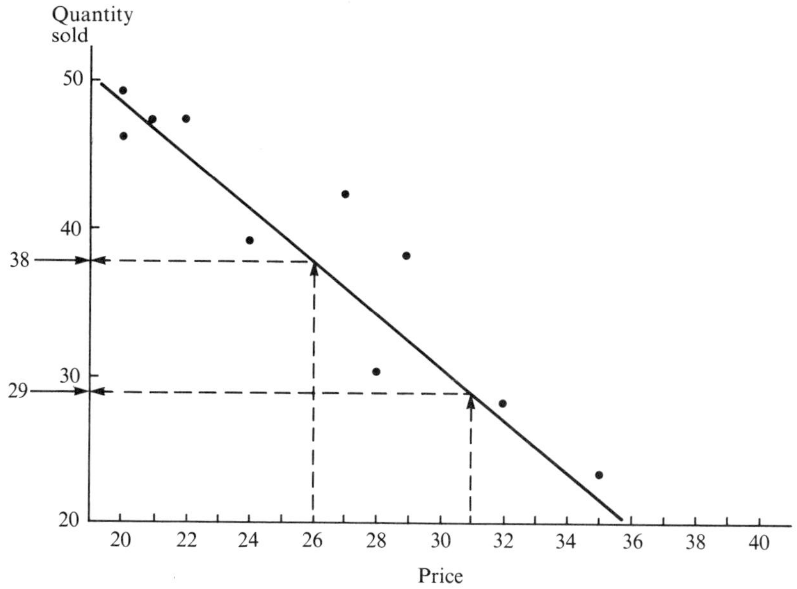

Figure 20.4

Scatter Diagram and Regression Line for the Data in Table 20.2

hand regression line in Figure 20.4. (*Note*: We added 38 + 46.8 because the slope in this problem is *negative*, −1.8. If the slope is *positive*, then we should *subtract* instead of add.)

The prediction equation may be written as follows:

$$\text{Quantity sold} = 84.8 - 1.8 \times \text{average selling price}$$

Thus, for an average price of 30¢ per dozen it is predicted that 30.8 dozen will be sold and for an average price of 25¢ per dozen, 39.8 dozen will be sold.

Compare Figures 20.2 and 20.4. Notice that the dots show a wider scatter around the regression line in Figure 20.4. This indicates that better predictions could be made for the Figure 20.2 problem than for the other problem. Figure 20.5(a) illustrates a scatter diagram where all the dots fall on a straight line. Perfect predictions could be made in such situations. However, this never occurs. Figure 20.5(b) illustrates a scatter diagram where the scatter of dots does not suggest any line. In such circumstances, it is not possible to construct a prediction equation. Parts (c) and (d) of this figure illustrate scatter diagrams where the suggested line of relationship is curved, not straight. Such regression curves are said to be *nonlinear* or *curvilinear* compared to *straight lines* which are said to be *linear*. When a curvilinear regression relationship is indicated, the foregoing procedure for determining a prediction equation does not apply. In such problems, determination of a prediction equation is best left to the statistician. When the relationship is curvilinear, the correlation coefficient is often called the *correlation index*.

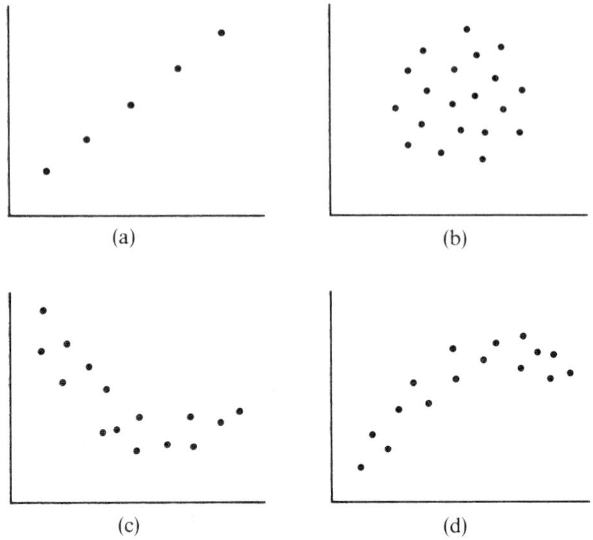

Figure 20.5

Scatter Diagrams Illustrating Various Types of Relationships

Review

Concepts and Terminology

Best-fitting regression line
Curvilinear regression
Deviation between observed and predicted values
Directly related
Dot chart
Freehand regression line
Height of regression line
Inversely related
Least squares prediction equation
Least squares regression line
Method of least squares
Negatively related
Negatively sloped
Nonlinear regression
Perfect prediction
Positively related
Positively sloped
Regression coefficient
Regression line
Scatter diagram
Slope of regression line
Square of the deviation
Straight-line regression
Sum of squares
Wider scatter

Problems

1. Should the predictor or predicted variable always be put on the horizontal scale in a scatter diagram?
2. Generally speaking, what does a regression line represent?

21

Association

A prediction equation, as we have seen, provides a basis for predicting the value of a variable based on specified values of a predictor variable. However, the *goodness of a prediction* depends on *how closely* the two variables are associated. Clearly, if the two variables are closely related, it is to be expected that the prediction will be better than if these variables bear little relationship to each other. We may assess the *degree of relationship* between two variables by inspecting the scatter of dots around the regression line. The tighter the scatter of dots (the closer the dots are to the regression line), the better and more useful the predictions made.

21.1 Coefficient of Determination

Suppose we have a random sample of weights of men aged 25 and compute the mean weight as 170 pounds. An additional man is to be selected from the population. What would you predict as his weight? If the only information available for the population of men is the sample of weight data, the best prediction we can make is that the man's weight is equal to the sample mean (170 pounds).

Of course, a man's weight is related to various factors such as height, heredity, type of work he does, how much he exercises, etc. If we also obtain information on the height of each man in the sample, we could construct a scatter diagram of weight and height and determine a prediction equation to predict weight based on height.

Using the prediction equation and assuming that we are told the height of the additional man selected, we could predict his weight as, perhaps, 185 pounds. This represents an improved prediction because it is based on more information (height and weight and the relationship between these two variables). Suppose we observe that this man actually weighs 190 pounds. Let us summarize these various weight measures.

1. Observed (actual) weight = 190 lb
2. Predicted weight from the prediction (regression) equation = 185 lb
3. Sample mean weight = 170 lb

Using these data, compute differences as follows:

(a) Observed weight − sample mean weight = 190 − 170
= 20 lb

(b) Predicted weight − sample mean weight = 185 − 170
= 15 lb

(c) Observed weight − predicted weight = 190 − 185
= 5 lb

The foregoing differences (a), (b), (c) provide a basis for determining *how closely* the two variables weight and height are associated and *how good* a prediction of a man's weight can be made based on his height. Let us first develop a few concepts.

The (a) difference is called the *total deviation* of an observed value from the sample mean. This difference, 20 pounds, can be partitioned into two parts. One part is the (b) difference (15 pounds) and the other part is the (c) difference (5 pounds).

The (b) difference is called the *deviation due to regression* (due to the relation of weight to height as expressed by the prediction equation). This difference (15 pounds) represents the improvement in the prediction over using the sample mean (170 pounds). In other words, taking into consideration that weight is related to height, we improve the first prediction of 170 pounds by adding 15 pounds to reflect the additional weight we would expect him to have due to his height (based on the prediction equation).

This implies that the sample mean weight must be adjusted up or down to predict the weight of a man based on his height. For a taller man, more than 15 pounds would have to be added; for a shorter man, less than 15 pounds would have to be added; and, for a very short man, a certain number of pounds would have to be subtracted from the sample mean weight. Clearly, it is to be expected that a tall man's weight will tend to be more than the average weight and a short man's weight will tend to be less than the average weight.

The (c) difference is called the *unexplained deviation* and represents the part of the man's weight which cannot be explained by how tall he is according to the prediction equation. This difference (5 pounds) is related to other factors which affect his weight (and that are not included in the regression analysis, such as heredity, etc.) The three differences are related as follows:

(a) difference = (b) difference + (c) difference

Or, we may write more technically

$$\text{Total deviation} = \text{deviation due to regression} + \text{unexplained deviation}$$

As noted in our discussion of the least squares prediction equation in the previous chapter, statisticians are usually concerned with the *squares of the deviations*. Suppose for each man in the sample we determine actual (observed) weight and predicted weight based on the prediction equation. Then, we obtain three sums, called *sums of squares*, as follows:

(a) *The total deviation* (*observed weight* − *sample mean weight*) is determined for each man in the sample, each deviation is squared, and the squared deviations are added for the men in the sample. This sum is called the *total variation* in the sample of weights.
(b) *The deviation due to regression* (*predicted weight* − *sample mean weight*) is computed for each man in the sample, each deviation is squared, and the squared deviations are added for the sample of men. This sum is called the *variation due to regression* or the *explained variation*. In other words, this represents the part of the total variation in the sample of weights that is *explained* by the variation in the men's heights (based on the regression equation).
(c) *The unexplained deviation* (*observed weight* − *predicted weight*) is computed for each man in the sample, each deviation is squared, and the squared deviations are added for the sample of men. This sum is called the *unexplained variation* and represents the part of the total variation in the sample of weights *not explained* by the variation in height. These three *sums of squared deviations* are related as follows:

$$\text{Explained variation} + \text{unexplained variation} = \text{total variation}$$

This is an important relationship in statistics. If each measure of variation in the foregoing relationship is divided by the *total variation* we obtain

$$\frac{\text{Explained variation}}{\text{Total variation}} + \frac{\text{unexplained variation}}{\text{total variation}} = 1$$

Or, in other words

$$\text{Proportion of the total variation explained} + \text{proportion of the total variation unexplained} = 1$$

The first proportion in this expression, the proportion of the total variation in the sample of weights explained by the variation in heights, is called the *coefficient of determination*. This coefficient represents a measure of *how closely* the two variables weight and height are associated and *how good* a prediction of a man's weight can be made based on his height.

If two variables such as weight and height are not related at all, as in Figure 20.5(b), *no part* of the variation in weight will be explained by or associated with the variation in height and the coefficient of determination will be equal to *zero*. Then, the variable height is worthless as a predictor of weight. On the other extreme, if weight and height are perfectly related, as in Figure 20.5(a), *all* of the variation in weight will be explained by or associated with the variation in height and the coefficient of determination will be equal to *unity* (1). Then, the variable height is a perfect predictor of weight.

Therefore, the lowest value the coefficient of determination can have is zero. This indicates that the two variables are not associated and that useful predictions of one variable based on the other cannot be made. Also, the highest value the coefficient of determination can have is unity (1). This indicates that the two variables are perfectly associated and perfect predictions of one variable can be made based on the other variable.

Most of the time, in practice, the coefficient of determination will have a value between 0 and 1. The closer this coefficient is to 0, the less the two variables are associated and the less useful is one as a predictor of the other. The closer this coefficient is to 1, the more closely the two variables are associated and the better is one as a predictor of the other.

21.2 Coefficient of Correlation

Usually, statisticians compute the *square root* of the coefficient of determination and call this the *coefficient of correlation.* (As noted in chapter 8, the square root of 25 is 5 because $5 \times 5 = 25$, the square root of 0.81 is 0.90 because $0.90 \times 0.90 = 0.81$, etc.). The lowest numerical value of the coefficient of correlation is 0, the highest numerical value is 1, and in most problems encountered in practice the numerical value of the coefficient of correlation is between 0 and 1. The correlation coefficient is an abstract number and can be interpreted by squaring (which results in the corresponding coefficient of determination).

For example, suppose the correlation coefficient for the variables "typing speed" and "number of months of experience" is computed as 0.80 for a sample of typists. This is a relatively high value for this coefficient because the highest possible value is 1.00. We may interpret this coefficient by squaring it: $0.80 \times 0.80 = 0.64$ which is the corresponding coefficient of determination. As previously discussed, 0.64 indicates that 64% of the variation in typing speed is associated with (or explained by) the variation in number of months of experience (based on the regression equation). We may also state that $1 - 0.64 = 0.36$ indicates that 36% of the variation in typing speed is not associated with (or not explained by) the number of months of experience, but is related to other factors like mechanical ability, interest in the work, etc.

We may also look at this problem the other way and note that 64% of the variation in months of experience is associated with the variation in typing speed and that 36% of the variation in months of experience is associated with other factors (based on the regression equation).

The coefficient of determination is always a positive number or zero; however, the coefficient of correlation may be zero, positive, or negative. If two variables are *positively related* (as discussed in the previous chapter), the correlation coefficient is *positive.* On the other hand, if the variables are *negatively related*, the coefficient is *negative*. For example, the correlation coefficient computed for "selling price" and "quantity sold" for a commodity may be −0.75. Notice that this coefficient is negative (if selling price *increases* quantity sold may be expected to *decrease*). Then, $-0.75 \times -0.75 = 0.5625$ or, more simply, 0.56 is the corresponding coefficient of determination. This indicates that 56% of the variation in one of these variables (say quantity sold) is associated with the variation in the other variable (selling price) and 44% of the variation ($1 - 0.56 = 0.44$) is associated with other factors (based on the appropriate regression equation). (**Note 21A**, at the end of the chapter, shows how to calculate the correlation coefficient.)

Review

Concepts and Terminology

Coefficient of correlation
Coefficient of determination
Degree of relationship
Deviation due to regression
Explained variation
Goodness of a prediction
Perfect association
Total deviation
Total variation
Unexplained deviation
Unexplained variation
Variation due to regression

Problems

1. The mean score on a mechanical ability test for a random sample of seniors is 220. What would you predict as the score obtained by an additional senior randomly selected from this population?
2. The following prediction equation was constructed for a random sample of 10 farms:

   Predicted yield (bushels per acre)
   $= 25 + 0.2 \times$ amount of fertilizer used (pounds per acre)

   The mean yield per acre for the sample is 45 bushels. Farm A, included in the sample, used 110 pounds of fertilizer and had a yield of 65 bushels per acre.
   (a) What is the predicted yield for a farm when 110 pounds of fertilizer are used?
   (b) Determine the total deviation, the deviation due to regression, and the unexplained deviation for farm A.
   (c) How do you interpret the deviation due to regression?
   (d) How do you interpret the unexplained deviation?
3. The following prediction equation was determined for a random sample of earthworms:

   Predicted length of earthworm (cm) $= 2.0 + 0.5 \times$ amount of soil intake (grams)

The mean length for the sample of earthworms is 25 centimeters. One of the earthworms in the sample had a recorded soil intake of 60 grams and measured 20 centimeters in length.
(a) Applying the prediction equation, what would you predict (or estimate) as the length of an earthworm with a recorded soil intake of 60 grams?
(b) Compute the total deviation, deviation due to regression, and the unexplained deviation for the specified earthworm in the sample.

4. A lawyer handling a tax case had a statistician construct an equation to predict the depreciated value of a type A item of equipment (as defined in his brief). The equation, which applies to type A items in use for a two-year period, is

$$\text{Depreciated value (\$)} = \$12{,}000 - \$15 \times \text{weight of item (lb)}$$

The sample of type A items used to provide the basic information for the analysis showed a mean depreciated value of $4,000. A specified sample item, which had an actual depreciated value of $2,000, was found to weigh 700 pounds.
(a) What is the predicted (or estimated) depreciated value of a type A item which weighs 700 pounds (after a two-year period of use)?
(b) Determine the total deviation, the deviation due to regression, and the unexplained deviation for the specified sample item.

5. A correlation analysis based on a random sample of 50 trees involved the variables "height of a tree (in feet)" and "age of a tree (years)." The variation in tree height explained by the relationship of height to age was computed as 600 and the unexplained variation in tree height was computed as 900.
(a) What is the total variation in tree height for the sample of 50 trees?
(b) Compute the coefficient of determination and explain what it means.
(c) Determine the coefficient of correlation for this analysis.

6. An analysis of the relationship between "crime rate" and "proportion of families with children aged 16 or under where both parents have full-time employment" was made for a random sample of 65 cities. The coefficient of correlation for these variables was computed as 0.86. Interpret this measure.

7. The following prediction equation was constructed for a department store for a random sample of 20 weeks: Predicted number of customer complaints = 2.50 + 0.05 × number of sales made. In terms of this illustration, explain how you would determine the—
(a) Total variation in number of customer complaints
(b) Explained variation in number of customer complaints
(c) Unexplained variation in number of customer complaints

8. In a correlation analysis involving the variables "number of insurance policies written during a month" and "number of TV spot advertisements run during a month," the total variation in "number of insurance policies written during a month" was computed as 2,000 and the variation in this variable not explained by the other variable was computed as 500 for a random sample of months (based on the regression equation).
(a) Determine the explained variation for "number of insurance policies written during a month."
(b) Determine the coefficient of determination and explain what it means.

9. The coefficient of correlation was determined as 0.70 for the two variables "research ability" and "writing ability." How do you interpret this coefficient?

10. The correlation coefficient computed for the two variables "quantity sold" and "price" is −0.95. How do you interpret this coefficient?

Note 21A

Computation of the Correlation Coefficient

$$\textit{Correlation coefficient} \qquad r = \frac{\sum xy}{\sqrt{\sum x^2 \times \sum y^2}}$$

where $\sum xy$ and $\sum x^2$ are computed as shown in **Note 20A**

$$\sum y^2 = \sum Y^2 - \frac{(\sum Y)^2}{n}$$

X and Y are the variables in the analysis

$\sum Y$ = sum of the Y observations

$\sum Y^2$ = sum of the squares of the Y observations

n = number of pairs of X and Y values

Referring to the sample data in **Note 20A**, we have the following:

$\sum x^2 = 84.67$ (from **Note 20A**)

$\sum xy = 328.67$ (from **Note 20A**)

Determine from the data in **Note 20A** that $n = 12$ and $\sum Y = 467$. If each Y value is squared ($33 \times 33 = 1089$, $45 \times 45 = 2028$, etc.) and the squares are added, we obtain $\sum Y^2 = 19{,}487$. Using this information, we compute

$$\sum y^2 = 19{,}487 - \frac{(467)^2}{12} = 1312.92$$

Finally,

$$r = \frac{328.67}{\sqrt{(84.67) \times (1312.92)}} = 0.99$$

This is a considerably higher value of r than is usually obtained in real problems. (We have artificially constructed this set of data.) Then, the coefficient of determination is $0.99 \times 0.99 = 0.98$ which indicates that 98% of the variation in one of the variables (say, typing speed) is associated with the variation in the other (say, experience) and only 2% ($1 - 0.98 = 0.02$) is associated with other related factors (based on the appropriate regression equation).

22

Further Concepts and Pitfalls

22.1 Logically Related Variables

The foregoing chapters on association and prediction present powerful statistical methods used in various fields of application. These methods provide a useful basis for evaluating the relationship between two variables and for predicting the value of one variable based on another. The reader is cautioned, however, to apply these methods only to problems where the two variables are *logically related*; otherwise, meaningless results will be obtained. The fact that the correlation coefficient computed for a pair of variables is high does not in itself signify a *cause and effect relationship* between the variables. This may be purely a *chance result* and have no logical basis whatsoever.

For example, say it is observed that the correlation coefficient computed for the number of passenger cars sold in Cleveland and the number of crimes committed in Nome, Alaska for a certain period is very high. Would anyone conclude that these variables are related or that it would be sensible to determine a prediction equation to predict the number of passenger cars sold in Cleveland based on the number of crimes committed in Nome, Alaska? Clearly, this would be a worthless exercise. Such computations are called *nonsense correlations.*

Suppose that for a period it is found that the average price of single family homes sold is highly correlated with the average retail price of men's suits. Should it be concluded that these variables are associated and that an increase in the average retail price of men's suits *causes* an increase in the average price of homes sold? Of course not. High correlations in such instances usually result from a *common underlying factor*. For example, both variables may be affected by an inflationary trend in the economy which pushes up prices generally, including the average prices of homes sold and men's suits. Such correlations are called *spurious correlations.*

The illustrations just presented are intended to emphasize that the methods of correlation and regression discussed in the previous chapters are *mechanical procedures* which may be applied to any pair of variables and that care must be taken that the variables included in an analysis are sensibly related if useful results are to be obtained.

22.2 Evaluation and Extrapolation

These methods are useful to *evaluate hypotheses* developed by subject-matter specialists. An agricultural specialist may hypothesize that the quality of soybean oil is dependent on the quantity of a certain chemical in the solvent used to extract the oil from the bean. This hypothesis may be evaluated by collecting data on the quality of soybean oil and the quantity of the chemical used for a number of specimens of soybean oil. Then, the correlation coefficient may be computed for soybean quality and chemical quantity and a prediction equation may be computed to predict quality of soybean oil based on the quantity of the chemical in the solvent. These results may be used to evaluate the hypothesis.

Or, it may be hypothesized that the score obtained by a job applicant on a social attitudes test is a useful predictor on how well the applicant will get along with the people he must deal with on the job. This hypothesis may be evaluated by collecting data on test scores for a sample of new employees and obtaining ratings from supervisors on how well these employees get along with the people they deal with in their work. Then, the coefficient of correlation could be computed for the two variables "test score" and "supervisor rating." A prediction equation to predict supervisor rating (which is used as a measure of getting along) based on test score could also be computed.

If some aspects of a problem under study are not adequately understood, an *exploratory correlation and regression analysis* may be useful. For example, what is the single most important factor related to the number of crimes committed by juveniles in a community? Is it the degree of family harmony prevailing in the community, average family income, or the average level of educational attainment of the heads of families? Such a study would require separate analysis based on sample data of the relationship between the number of crimes committed and a measure of each of the possible related factors. Of course, exploratory analyses are risky since it is always possible that a particular relationship looks acceptable only because of the vagaries of random sampling. Hence, it is necessary to carefully evaluate a pair of variables to determine whether it is logical to expect them to bear a relationship to each other.

Prediction equations are often *extrapolated* when predictions are made. Suppose a business analyst computes an equation to predict family expenditures for food based on family income, where family incomes in the analysis range from a low of $5,000 a year to a high of $20,000 a year. The range, $5,000–$20,000, is called the *observed range* of the data. If food expenditures are predicted for a family with $30,000 a year income using this equation, the analyst is extrapolating the line of relationship implied by the equation (the regression line) to make the prediction. *Extrapolation* means going outside the limits of the observed range. In our illustration, $30,000 a year income is outside of the $5,000–$20,000 range of the data.

It is always risky to make a prediction when extrapolation is involved. For example, the prediction equation in our illustration may show that food expenditure increases as family income increases. However, would you expect this relationship to hold for higher and higher family incomes? Clearly, there is a limit to how much a family can eat. We would expect that at some income level food expenditures would no longer increase because the family is spending as much as it requires for food and will now use additional income to increase other types of expenditures, such as for entertainment, education, or investment. Whenever extrapolation is involved, the analyst must evaluate his problem carefully to determine whether the prediction equation is suitable for prediction.

22.3 Analyses Using More Than Two Variables

In our discussion of correlation and prediction we were concerned with only two variables. Often, three or more variables are involved. For example, it may be desired to compute an equation to predict the annual sales of washing machines based on the average annual price of washing machines, average annual income, and the average annual amount of savings. In this problem, we have four variables, the predicted variable is "annual sales of washing machines" and the other variables represent three predictor variables.

Problems which involve three or more variables are to be handled by the *methods of multiple correlation and regression.* This is a more complicated area of application of statistical methods and is not suitable for discussion in this text.

Review

Concepts and Terminology

Exploratory analysis
Extrapolation
Logically related variables
Multiple correlation
Multiple regression
Nonsense correlation
Observed range
Spurious correlation

Problems

1. If the coefficient of correlation computed for a pair of variables is close to one, does this prove that there is a cause and effect relationship between these variables? Defend your answer.
2. Illustrate and explain the meaning of nonsense correlation, spurious correlation.

3. During an interval of a few months, an analyst found that the number of accidents in manufacturing plants was highly correlated with the number of inches of rainfall in a nearby city. The analyst concluded that the number of accidents in manufacturing plants could be predicted based on the number of inches of rainfall in the nearby city. Comment.

4. Illustrate an exploratory correlation and regression analysis.

5. The following prediction equation was computed based on a sample of 100 medium-sized farms: Predicted yield (bushels per acre) = 15 + 0.3 × amount of fertilizer used (pounds per acre). The sample of farms used in this analysis applied between 50 and 80 pounds of fertilizer per acre. Using this equation, what would you predict as the yield if 120 pounds of fertilizer were applied per acre? Comment on the usefulness of this predicted yield.

6. The number of deaths of persons past 65 years of age and the number of infant deaths were collected for a sample of 50 medium-sized cities for a given week. These two mortality measures were found to have a correlation coefficient of 0.85. Would it be useful to compute a regression equation to predict infant deaths based on deaths of persons past 65 years of age? Why?

7. It was found that during a period of rising incomes and general prosperity, the correlation coefficient computed between salaries paid to policemen and liquor consumption was 0.95. Is it useful to compute a regression equation to predict liquor consumption based on policemen salaries? Why?

8. Suppose you want to compute a regression equation to predict the height of a certain type of plant after a month of growth, based on a sample of 50 plants. Which of the following would make the best selection as the predictor variable—amount of fertilizer used, age of the person who planted the seed, day of the week the seed was planted? Why?

9. Suppose you want to determine a regression equation to predict the number of single-family homes sold based on data collected from a sample of 30 cities. Which of the following would be the best predictor variable to use—the number of auto accidents, the number of crimes committed, the number of marriages the year before? Why?

10. The following equation was computed based on a sample of 45 banks:

$$\text{Predicted number of new loans made} = 25 + 0.07 \times \text{number of new depositors the month before}$$

The banks in this sample had from 25 to 60 new depositors the month before. Using this equation, what would you predict as the number of new loans made for a bank which had 90 new depositors the month before? Would you consider it safe to use this prediction?

Appendix A

ADDITIONAL METHODS

A.1

The *t* Distribution

In **Note 13B** at the end of chapter 13 it was indicated that the statistic

$$z = \frac{\bar{X} - \mu}{\frac{\sigma}{\sqrt{n}}}$$

is distributed approximately like the normal distribution. The approximation becomes better as the sample size gets larger. It was noted that the sample size must be at least 30 for the normal curve to provide a sufficiently good approximation. As previously noted, if the population from which the sample is selected is itself distributed approximately like the normal curve, then the sample size could be less than 30 for the normal curve to provide a useful approximation.

In the typical sampling situation the population standard deviation σ is unknown and the sample standard deviation s is used as an estimate of σ. Then, the z statistic is computed as follows:

$$z = \frac{\bar{X} - \mu}{\frac{s}{\sqrt{n}}}$$

In this form, the distribution of the z statistic is approximately like the normal distribution only if the sample size is 30 or more. If the sample size is less than 30 and the distribution of the population from which the sample was selected is similar to the normal curve distribution, then we call this the *t statistic*, because it is distributed like the *t distribution*. We may write the t statistic as follows:

$$t = \frac{\bar{X} - \mu}{\frac{s}{\sqrt{n}}}$$

Actually, for any sample size n, the t statistic has the t distribution if the population distribution is approximately like the normal curve. The shape of the t distribution depends on the number of degrees of freedom (df) involved in a problem. The number of df for the t statistic as presented in the foregoing equation is sample size $-$ 1 (or $n - 1$). In some problems the t statistic is computed in a different way and df is differently determined. This is illustrated in Section A.2.

The t statistic is also called *Student's t*, since it was introduced by W. S. Gosset who published his findings under the pen-name "Student." Figure A.1.1 presents a comparison of the t distribution and the normal (z) distribution. Notice that both distributions are symmetrical around the mean ($= 0$) and each has a central peak. The t distribution is not as high as the normal distribution in the central portion, but it is higher in the tails, so the t distribution is more spread out. Therefore, a smaller proportion of the area under the curve is concentrated around the mean for the t distribution than for the normal distribution and a larger proportion is in the tails. Consequently, if the $\pm z$ limits shown in Figure A.1.1 mark off a specified proportion of the area under the normal curve, the corresponding t limits must be further in the tails of the t distribution to mark off an equal proportion of the total area under the t distribution curve. This relationship between the z limits and the t limits is shown in the figure.

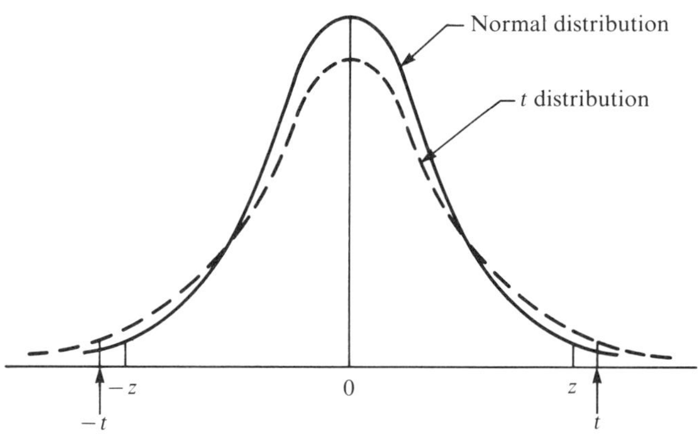

Figure A.1.1

Comparison of a t Distribution and the Normal Distribution

It is useful to compare the t values for various df with the corresponding z values. If the z values in Figure A.1.1 are ± 1.96, then the area under the normal curve between these limits is 0.95 and the area in both tails combined is $1.00 - 0.95$ or 0.05. If the t values in Figure A.1.1 are ± 1.98 for a t distribution with df $= 120$, the area under the t curve between these limits is also 0.95 and the area in both tails combined is 0.05. Notice that these t limits (± 1.98) are nearly equal to the corresponding z limits (± 1.96). However, the t limits required to include 0.95 of the area under the t curve increases in magnitude (moves further into the tails) as the number of degrees of freedom is reduced. For example, the corresponding t limits are ± 2.00 for df $= 60$, ± 2.04 for df $= 30$, and ± 2.23 for df $= 10$. Clearly, for large samples (n greater than or equal to 30) where df are large (29 or more), the t limits are not too different from the corresponding z limits. In other words, for large samples, the t distribution is sufficiently similar to the normal distribution so that the differences between the two distributions may be ignored.

Table B.5 in Appendix B presents limits for t distributions corresponding to different numbers of degrees of freedom and for a variety of areas included between the limits. The first column shows df from 1 through 29. The symbol ∞ denotes an infinite number of degrees of freedom.

The other column headings specify the area in *both tails* of the t distribution combined. These areas represent probabilities, as in the case of the normal distribution. The first probability (or area) column is "0.5," so the t values shown in this column cut off, *in both tails combined*, 0.5 of the area under the t curve (half of 0.5 or 0.25 in each tail). Referring to the probability column, "0.05," we note that for 10 df the t limits are ± 2.228. Therefore, $1 - 0.05$ or 0.95 of the area under the t curve for 10 df is included between $t = -2.228$ and $t = 2.228$ and 0.05 is the area in both tails combined.

It has already been observed that, as df (or sample size) becomes larger, the t distribution becomes more and more like the normal distribution. In the limit, with df infinite, the two distributions coincide. Notice that for df $= \infty$ (infinity), the t values are exactly equal to the corresponding z values. As an example, for 0.05 of the area in both tails (or 0.025 in each tail), $t = z = 1.96$.

A.2

Statistical Methods for Small Samples

In many applications of statistical methods, when small samples (sample size less than 30) are used and the population from which the sample was selected has a distribution similar to the normal distribution, the appropriate sampling distribution can be approximated by the t distribution and not the normal distribution.

Note 14A, at the end of chapter 14, shows how to construct a *confidence interval for a population mean* for large samples ($n = 30$ or more). A similar procedure may be used where the sample size is less than 30 and the population distribution can be approximated by the normal curve. However, the equation for the maximum expected error must be changed by substituting t for z as follows:

$$m = t \times \frac{s}{\sqrt{n}} \times \sqrt{\frac{N - n}{N - 1}}$$

Then, the confidence limits and confidence interval are determined as explained in **Note 14A**. The number of df $= n - 1$.

Suppose the sample in a problem is of size 20 and the confidence level is specified as 0.95. The number of df is $20 - 1 = 19$. Then, referring to Table B.5, find 19 in the "df" column and move across that row to the column headed ".05" (since $1 - 0.95 = 0.05$) and read out 2.093. This is the t value to enter in the foregoing equation for maximum expected error.

Note 16B, at the end of chapter 16, shows how to construct a *test of hypothesis relating to a population mean* based on a large sample. A similar test may be conducted when the sample size is less than 30 and the population distribution can be approximated by the normal curve. However, the critical value equation must be changed by substituting t for z as follows:

$$\text{Critical value} = t \times \frac{s}{\sqrt{n}} + \mu$$

Then, the test is conducted as explained in **Note 16B**. The number of df $= n - 1$.

Suppose the sample in a problem is of size 25 and the level of significance of the test is 0.01. Then, for a one-tail test, the appropriate t value is found in the column headed ".02" which represents a t value which cuts off 0.02 (or 2%) of the total area under the t distribution curve in both tails and 0.01 (or 1%) in one tail. The number of df = 25 − 1 = 24. The appropriate t value read out of Table B.5 is 2.492. If the test is a lower-tail test, use $t = -2.492$ and if an upper-tail test, use $t = 2.492$.

If the test is a two-tail test, then look for the t value in the column headed ".01" for df = 24 because we want a t value which cuts off 0.01 in both tails combined. This t value is 2.797.

Note 17B, at the end of chapter 17, shows how to construct a *test of hypothesis relating to the comparison of two population means based on large independent samples.* A similar test may be conducted where the samples are less than 30, each population is distributed approximately like the normal curve, and the *population variances are equal.* (The procedure described in **Note 17E** can be used to test whether two population variances may be considered to be equal.) However, the critical value must be computed as follows:

$$\text{Critical value} = t \times \sqrt{s^2 \times \left(\frac{1}{n_1} + \frac{1}{n_2}\right)}$$

where s^2 is the *pooled variance estimate* of the variance common to the two populations and is computed as follows:

$$s^2 = \frac{(n_1 - 1) \times s_1^2 + (n_2 - 1) \times s_2^2}{n_1 + n_2 - 2}$$

Then, the test is conducted as explained in **Note 17B**. The number of df $= n_1 + n_2 - 2$. The appropriate t value to use in a test is determined for a one-tail test or a two-tail test as just explained for tests discussed in **Note 16B**.

Note 17C shows how to construct a test of hypothesis relating to the comparison of two population means based on two large *dependent samples.* A similar test may be conducted where the number of sample *pairs* is less than 30 and the population of pair differences is considered to have a distribution which is approximately like the normal distribution. However, the critical value equation must be changed by substituting t for z as follows:

$$\text{Critical value} = t \times \frac{s}{\sqrt{n}}$$

Then, the test is constructed as explained in **Note 17C**. The number of df $= n - 1$, where $n =$ the number of sample pairs. The appropriate t value to use in a test is determined for a one-tail or a two-tail test as previously explained for tests discussed in **Note 16B**.

A.3

Nonparametric Tests

For the most part, statistical test procedures so far presented are applicable only if certain conditions pertaining to the population or populations involved are statisfied. For example, a condition may be that the population is at least approximately normal or that a parameter (such as the standard deviation) is equal for two populations. Tests of this type are called *parametric tests*. Another type of statistical test, useful in a variety of situations, is the *nonparametric test*. Such tests, generally, are applicable to a wider group of test situations since they impose fewer and looser restrictions.

Nonparametric tests involve simpler computations than parametric tests and are easier to apply and understand. Nonparametric tests are also called *distribution-free tests* since they are applicable in situations where the population or populations involved have unknown distributions. An important characteristic of nonparametric tests is their applicability to ranking data, ordinal-scale data, and nominal-scale or classificatory data. (These types of data were discussed in Section 4.2.) Such data cannot be tested by parametric test procedures. For example, nonparametric tests may be used to compare the ranking of ten applicants on suitability for a certain job by two interviewers, the preferences of teenage girls relating to certain cosmetics in two high schools, and the distributions by type of job held by employed females aged 20 to 30 in two cities.

We have already encountered an important group of nonparametric tests, the chi-square tests presented in chapter 18. We will consider certain additional nonparametric tests in Sections A.4 through A.7 of this appendix.

A serious weakness of nonparametric tests is that they are generally wasteful of information. That is, they do not always use all the information in the sample data. For example, if a sample of weights is used in a nonparametric test, only the weight rankings, not the weights themselves, may be used in the test. Often, a larger sample is required when a nonparametric test is applied than would be used if a comparable parametric test were applied instead.

Finally, in many situations a nonparametric test is used even if a suitable parametric test is available because the nonparametric test is quicker and easier to apply or because there is not enough assurance that the conditions relating to the parametric test are adequately met. Nevertheless, a parametric test is usually preferable to a nonparametric test, if it is applicable to a test situation.

A.4

Sign Test for Paired Data

One of the simplest nonparametric tests to apply is the *sign test*, which is based on the direction (sign) of a difference for a pair of observations. Suppose a sample of 24 drivers recorded the number of miles per gallon obtained when using a certain gasoline and then were asked to drive the same cars and determine the number of miles per gallon obtained when using the same gasoline with an additive introduced. Assume 16 drivers reported increased mileage (+ differences), 4 drivers reported reduced mileage (− differences), and 4 drivers reported no difference in mileage. Would you conclude, at the 0.05 level of significance, that the additive increased gasoline mileage?

In conducting the sign test, ignore the cases with no change (zero difference). Therefore, we are concerned with 16 pluses and 4 minuses and a sample size of $n = 20$. If the additive had no effect on mileage, we would expect that the mileage would be about the same with or without the additive and would differ for each driver solely because of chance. Then, we would expect that the number of pluses would be about the same as the number minuses. Hence, formulate the hypotheses:

1. *Hypothesis to be tested.* The proportion of pluses is 0.5 (or the additive has no effect on mileage)
2. *Alternative hypothesis.* The proportion of pluses is greater than 0.5 (or the additive increases mileage)

The decision rule is based on the sampling distribution of a proportion. The critical value may be computed as follows:

$$\text{Critical value} = z \times \sqrt{n \times p \times (1 - p)} + (n \times p)$$

where z = the z value (obtained from Table B.1) which cuts off an area under the normal curve in the lower tail, upper tail, or both tails combined (half in each tail) as determined by the alternative hypothesis and the level of significance of the test

n = sample size (the number of + or − differences)

p = the proportion of +'s as stated in the null hypothesis

In our problem, $z = 1.64$ (which cuts off 0.05 of the area under the normal curve in the upper tail), $n = 20$, and $p = 0.5$. The critical value is computed according to the foregoing equation:

$$\text{Critical value} = 1.64 \times \sqrt{20 \times 0.5 \times (1 - 0.5)} + (20 \times 0.5)$$
$$= 13.67$$

Decision rule If the number of sample observations in the category of interest (the number of pluses) is greater than 13.67, reject the hypothesis tested, accept the alternative hypothesis, and conclude that the additive increases mileage. Otherwise, conclude that the additive has no effect on mileage.

There is a technicality to be observed when applying this decision rule. The number of pluses is a discrete variable (as discussed in Section 4.1). The sampling distribution on which the decision rule is constructed is approximated by the normal distribution, which is a continuous distribution with a continuous variable. Therefore, the discrete variable "number of pluses" must be *corrected for continuity* (which means that it is converted to a continuous variable). This is accomplished by replacing the sample value of this variable by an interval. For example, the number of pluses in our problem is 16, and it is replaced by the interval 15.5–16.5. If the number of pluses were 10, the interval would be 9.5–10.5, etc.

Then, compare the upper limit of the interval with the critical value in a test if it is a lower-tail critical value. If comparison is made with an upper-tail critical value, the lower limit of the interval should be used.

The decision rule in our problem has the critical value 13.67, which is an upper-tail critical value (notice the alternative hypothesis). The number of pluses in our problem is 16 and the lower limit of the 15.5–16.5 interval is 15.5, which falls in the region of rejection. Therefore, we reject the null hypothesis and conclude that the additive increases mileage.

The effect of the additive on gasoline mileage could also be tested statistically by the parametric test procedure for small dependent samples discussed in Section A.2. This test is based on the *amount of difference* in mileage for the gasoline with and without the additive for the $n = 24$ drivers. This test is applicable only if the mileage differences are considered to be approximately normal in distribution.

A.5

Mann-Whitney U Test for Independent Samples

It is claimed that entrance salaries paid to programmers in city A are the same as in city B, on the average. Table A.5.1 presents the entrance salaries paid to a random sample of $n_1 = 10$ city A programmers and a random sample of $n_2 = 12$ city B programmers. Would you accept the claim based on these data? Use the 0.05 level of significance.

The *Mann-Whitney U Test* provides a test of the difference between means based on *rank sums*. The mean for the n_1 sample is \$5,100 and for the n_2 sample it is \$4466.67. Do these means result from a difference in population means or do they differ merely due to chance? Combining the two samples into one of size $n = n_1 + n_2$ or $10 + 12 = 22$, rank the salaries from lowest (rank 1) to highest (rank 22). These ranks are shown in Table A.5.1. The n_1 sample ranks total $R_1 = 136$ and the n_2 sample ranks total $R_2 = 117$.

If the population distributions of entrance salaries for city A and city B are the same, we would expect that for samples of equal size the rank sums would be about equal. However, if the mean salary is lower for one of the populations, we would expect that the sample ranks and the rank sum would tend to be smaller for that population. In our illustration, the rank sum R_2 (for city B programmers) is smaller than R_1 even though it is based on a larger sample. This does not appear to support the claim that, on the average, city A salaries are the same as city B salaries. However, let us conduct the test to determine whether the sample differences reflect only chance effects.

The Mann-Whitney test is based on the *U statistic* which is computed as

$$U = n_1 \times n_2 + \frac{n_1 \times (n_1 + 1)}{2} - R_1$$

Formulate the hypotheses:

1. *Hypothesis to be tested.* $\mu_1 = \mu_2$ (The populations of city A and city B entrance salaries are identical in distribution, so the means of the city A salaries μ_1 and city B salaries μ_2 are equal.)
2. *Alternative hypothesis.* μ_1 is not equal to μ_2

Table A.5.1

Entrance Salaries Paid to a Sample of $n_1 = 10$ City A Programmers and a Sample of $n_2 = 12$ City B Programmers—Ranks and Rank Sums for the Mann-Whitney U Test

Sample		Ranks Based on Combined Sample $n = n_1 + n_2$	
n_1	n_2	n_1	n_2
5,000	6,300	13	22
3,500	3,200	4	2
4,800	4,300	11	8
6,000	5,600	20	17
4,500	4,200	9	7
5,500	5,900	16	19
5,100	3,000	14	1
4,600	3,600	10	5
6,200	3,800	21	6
5,800	3,400	18	3
	4,900		12
	5,400		15
	Totals	136	117

If the hypothesis to be tested is true, then the combined sample of n observations may be considered to be one sample selected from one population. If n_1 and n_2 are each at least of size 10 then the sampling distribution of U is approximately like the normal curve.

The critical value is computed as follows:

$$\text{Critical value} = z \times \sqrt{\frac{n_1 \times n_2 \times (n_1 + n_2 + 1)}{12}} + \frac{n_1 \times n_2}{2}$$

where z = the z value (obtained from Table B.1) which cuts off an area under the normal curve in the lower tail, upper tail, or both tails combined (half in each tail) as determined by the alternative hypothesis and the significance level of the test.

n_1 = size of the first sample

n_2 = size of the second sample

In our problem $z = \pm 1.96$ (which cuts off 0.025 of the area under the normal curve in each tail and 0.05 in both tails combined), $n_1 = 10$, and $n_2 = 12$. The critical values are computed according to the foregoing equation:

$$\begin{pmatrix}\text{Lower-tail}\\ \text{critical value}\end{pmatrix} = -1.96 \times \sqrt{\frac{10 \times 12 \times (10 + 12 + 1)}{12}} + \frac{10 \times 12}{2} = 30.27$$

$$\begin{pmatrix}\text{Upper-tail}\\ \text{critical value}\end{pmatrix} = 1.96 \times \sqrt{\frac{10 \times 12 \times (10 + 12 + 1)}{12}} + \frac{10 \times 12}{2} = 89.73$$

Decision rule If U computed from the sample data is either less than 30.27 or more than 89.73, reject the hypothesis tested, accept the alternative hypothesis, and conclude that mean entrance salaries are not the same for city A and city B programmers. Otherwise, conclude that the mean entrance salaries are the same.

Applying the sample data and the equation for U previously presented, compute

$$U = 10 \times 12 + \frac{10 \times (10 + 1)}{2} - 136 = 39$$

Then, $U = 39$ falls between the two critical values and so lies in the region of acceptance. We conclude that mean entrance salaries for city A and city B programmers are the same.

The Mann-Whitney test is not as wasteful of information as the sign test, since it uses the ranking information contained in the data. If the combined sample of n observations has ties, ranks should be assigned to the tied observations as if they can be ranked. Then assign the mean of the ranks so obtained to each of the tied observations. For example, suppose the n observations are 20, 22, 22, 30, 34, 34, 34, 40. Assigning ranks, we have 1, 2, 3, 4, 5, 6, 7, 8. Then, the two observations of 22 have ranks 2 and 3 which have the mean $(2 + 3)/2 = 2.5$, and the three observations of 34 have ranks 5, 6, 7 which have the mean $(5 + 6 + 7)/3 = 6$. Then, the ranking becomes 1, 2.5, 2.5, 4, 6, 6, 6, 8.

If samples n_1 and n_2 are not large enough to permit the use of the normal distribution in conducting the test, special tables will have to be used, such as those in *Nonparametric Statistics for the Behavioral Sciences*, Sidney Siegel, 1956, McGraw-Hill Book Company, Inc.

A. 6

Wilcoxon Test for Paired Data

The *Wilcoxon rank-sum test for paired data* is more powerful than the sign test (Section A.4) because it takes into account the *magnitudes* of pair differences as well as the *signs*.

Table A.6.1 presents the heights of pairs of plants for 15 varieties of the plant where one of each pair was treated according to method A and the other according to method B. Would you conclude, at the 0.01 level of significance, that the distribution of plant heights is the same for both methods?

The Wilcoxon test begins by computing the pair differences in height and ranking these differences according to *amount of difference without regard to sign*. The smallest *difference magnitude* is assigned rank 1. In Table A.6.1, the smallest difference magnitude, 2, is assigned rank 1, the next higher difference magnitude, 4, is assigned rank 2, the third smallest difference magnitude, 5, is assigned rank 3, etc. (Treat ties as explained in Section A.5.) These ranks are shown in the table separately for the positive differences ($+d$) and the negative differences ($-d$). Zero differences are ignored, so the sample size involved in a test is n = *the number of pairs with nonzero differences*. The rank sums of the $+d$ and $-d$ ranks are also presented in Table A.6.1. Clearly, if the two methods A and B have the same effect on plant height, we would expect the number of ranks as well as the rank sums for $+d$ and $-d$ to be about equal. On the other hand, if one of the methods produced taller plants, say method A, we would expect that more than half the A − B differences would be $+d$'s and the rank sum for $+d$ would exceed the rank sum for $-d$.

The Wilcoxon test is based on the T *statistic* which is the *smaller of the two rank sums* ($+d$ sum or $-d$ sum). In our problem $T = 8$, the $-d$ sum as shown in Table A.6.1. When the number of pairs n is say 10 or more (25 or more is better), the T statistic is approximately like the normal curve in distribution. Formulate the hypotheses:

1. *Hypothesis to be tested.* Methods A and B have the same effect on plant height
2. *Alternative hypothesis.* These methods have different effects

Table A.6.1

Heights of 15 Varieties of a Plant for method A and method B: Pair Differences, Ranks, and Rank sums for the Wilcoxon Test

Height (cm)		Difference	Rank	
A	B	$d = A - B$	$+d$	$-d$
40	35	5	3	
15	17	−2		1
30	22	8	6.5	
20	14	6	4	
45	35	10	9	
50	30	20	15	
42	31	11	10	
36	43	−7		5
25	29	−4		2
48	36	12	11	
20	12	8	6.5	
18	9	9	8	
37	24	13	12	
48	32	16	13	
50	32	18	14	
		Totals	112	8

The critical value is computed as follows:

$$\text{Critical value} = z \times \sqrt{\frac{n \times (n + 1) \times (2n + 1)}{24}} + \frac{n \times (n + 1)}{4}$$

where $z =$ the z value (obtained from Table B.1) which cuts off an area under the normal curve in the lower tail, upper tail, or both tails combined (half in each tail) as determined by the alternative hypothesis and the significance level of the test

$n =$ the number of pairs with nonzero differences

In our problem, $z = \pm 2.58$ (which cuts off 0.005 of the area under the normal curve in each tail and 0.01 in both tails combined) and $n = 15$. Critical values are computed according to the foregoing equation, as follows:

$$\begin{pmatrix}\text{Lower-tail}\\ \text{critical value}\end{pmatrix} = -2.58 \times \sqrt{\frac{15 \times 16 \times 31}{24}} + \frac{15 \times 16}{4} = 14.57$$

$$\begin{pmatrix}\text{Upper-tail}\\ \text{critical value}\end{pmatrix} = 2.58 \times \sqrt{\frac{15 \times 16 \times 31}{24}} + \frac{15 \times 16}{4} = 105.43$$

Decision rule If the T statistic determined from the sample data is either less than 14.57 or more than 105.43, reject the hypothesis tested, accept the alternative hypothesis, and conclude that the distribution of plant heights is not the same for methods A and B. Otherwise, conclude that the distributions are the same.

The T statistic $= 8$ for our problem (the smaller rank sum) which is below the lower-tail critical value and so falls in the region of rejection. Therefore, we conclude that the distribution of plant heights is not the same for methods A and B. Special tables are required if n is small, such as the tables in Siegel (noted in Section A.5).

A. 7

Correlation of Ranks

Suppose ten typists are ranked on experience X (1 = most experienced) and number of typing errors made Y (1 = largest number of errors) as shown in Table A.7.1. We could compute the correlation coefficient using the approach of **Note 21A**. However, a more rapid method when *ranks* are involved is to compute r', the *correlation of ranks*, as follows:

$$r' = 1 - \frac{6 \times \Sigma D^2}{n \times (n^2 - 1)}$$

where D = the difference between ranks
n = number of pairs of ranks
ΣD^2 = sum of the squares of the D's for the n pairs of ranks

Applying this equation to our problem (Table A.7.1), we obtain

$$r' = 1 - \frac{6 \times (310)}{10 \times (100 - 1)} = -0.88$$

This is identically the same result which would be obtained by using the method of **Note 21A** and indicates that the greater the experience (X), the fewer are the errors (Y).

When it is desired to determine the correlation between two variables and a rapid approximation to r is satisfactory, a useful approach is to rank the X and Y observations separately and compute the correlation of ranks r' based on the ranks. When data are ranked and ties occur, ranks should be assigned by assuming that the tied data can be ranked and then assigning the mean of the ranks so obtained to each of the tied values. For example, say we have the observations

Table A.7.1

Ten Typists Ranked on Experience (X) and Number of Errors Made (Y)

Typist	X	Y	$D = Y - X$	D^2
A	6	3	−3	9
B	7	5	−2	4
C	8	1	−7	49
D	9	4	−5	25
E	10	2	−8	64
F	1	10	9	81
G	2	8	6	36
H	3	7	4	16
I	4	9	5	25
J	5	6	1	1
			Total	310

20, 25, 28, 28, 29, 31, 31, 31, 33. Start by assigning ranks 1, 2, 3, 4, 5, 6, 7, 8, 9. Note that rank 3 is assigned to the first 28 and 4 to the other. Then, computing the mean, $1/2(3 + 4) = 3.5$, assign the rank 3.5 to each observation of 28. Also, computing the mean, $1/3(6 + 7 + 8) = 7$, assign the rank 7 to each observation of 31. Then, the ranks for these observations are 1, 2, 3.5, 3.5, 5, 7, 7, 7, 9.

Problems

1. A random sample of 16 college juniors selected from the colleges in a region showed a mean of \$650 earned last summer, with a standard deviation of \$50. Construct a 0.98 confidence interval for the mean earnings last summer for the 500 juniors in the colleges in the region. Assume that the summertime earnings are distributed like the normal curve.
2. A random sample of 25 earthworms collected in a field showed a mean length of 2.75 inches, with a standard deviation of 0.75 inches. Construct a 95% confidence interval for the mean length of earthworms in the field. Ignore the fpc in your computations. Why should the fpc be ignored? Assume that earthworm lengths have a normal distribution.
3. The president of a large law firm claims that it takes a mean of 15 days to prepare legal briefs for the type of cases the firm handles. The senior partner believes that this claim is too high. A random sample of 12 cases handled by the firm last year showed a mean of 13 days were required to prepare briefs, with a standard deviation of 3.5 days. Would you conclude, at the 0.05 significance level, that the president's claim should be accepted? Assume that the time required to prepare legal briefs has a normal distribution. (State the appropriate hypotheses and the decision rule.)
4. It is claimed by a horticulture firm that type G seed will produce plants with a mean height of 8 centimeters after a certain period of growth. A random sample of 20 seeds resulted in plants with a mean height of 9.2 centimeters and a standard deviation of 1.35 centimeters. Assuming that plant heights are distributed like the normal curve, would you accept or

reject the claim? Use the 0.02 significance level and conduct a two-tail test. (Show the appropriate hypotheses and the decision rule.)

5. It is claimed that the time required to accomplish a task is the same for workers with five or more years of experience as for workers with two to four years of experience. A random sample of 10 workers with five or more years of experience took a mean time of 15.2 minutes to accomplish the task (with a standard deviation of 3 minutes) compared with a mean of 18.7 minutes (with a standard deviation of 4 minutes) for an independent random sample of 12 workers with two to four years of experience. Assuming that the time required is distributed like the normal distribution for both categories of workers, with equal population variances, would you accept the claim? Use the 0.01 level of significance. (Show the appropriate hypotheses and the decision rule.)

6. A sample of 12 high-school boys ran a certain distance in a mean of 50 minutes on a dry day compared with a mean of 62 minutes on a rainy day. The difference between running times (rainy-day time − dry-day time) was computed for each runner and was found to have a mean of 12 minutes and a standard deviation of 4.5 minutes. Assuming that the population of differences is distributed like the normal curve, would you conclude that the mean running time is the same for dry days and rainy days? Use the 0.05 significance level. (Show the appropriate hypotheses and the decision rule.)

7. A sample of 26 matched pairs of guinea pigs was used in a drug experiment. One guinea pig of each pair was randomly selected for sample A and the other guinea pig assigned to sample B. Type X5 of a drug was injected into the sample A animals and type X8 of the drug was injected into the sample B animals. The time required for the drug to take effect (in minutes) was recorded for each guinea pig, as follows:

Pair	*Sample A*	*Sample B*	*Pair*	*Sample A*	*Sample B*
1	4	3	14	6	7
2	3	5	15	3	2
3	5	6	16	5	4
4	4	4	17	4	3
5	5	7	18	5	6
6	3	6	19	3	2
7	3	2	20	6	4
8	5	4	21	4	3
9	6	6	22	3	2
10	4	5	23	3	5
11	6	4	24	6	4
12	5	3	25	6	5
13	3	5	26	5	3

Using the 0.05 level of significance, would you conclude that the time required for the drug to take effect is the same for types X5 and X8? Conduct a sign test. (Show the appropriate hypotheses and state the decision rule.)

8. In Problem 7, conduct the test because you believe that the time required for the drug to take effect is more for type X5.

9. A newly developed analytical reasoning test was administered to a random sample of $n_1 = 14$ seniors in college A and a random sample of $n_2 = 12$ seniors in college B. The following test scores were obtained:

College A	College B
113	123
105	122
115	126
128	107
135	120
102	139
137	142
130	110
138	141
153	144
156	150
155	146
160	
158	

Would you conclude that the mean score on this test is the same for seniors in the two colleges? Use the 0.02 level of significance and conduct a Mann-Whitney U test. (Show the appropriate hypotheses and state the decision rule.)

10. Refer to Problem 9. Reverse n_1 and n_2 so that the previous n_1 sample is now the n_2 sample and the previous n_2 sample is now the n_1 sample. Conduct the test and compare the result with that obtained in Problem 9. Does it matter which of the two samples is called n_1?

11. A random sample of 18 workers was selected and each worker was required to accomplish a specified task using a domestic machine. Then, each worker was required to accomplish the same task using an imported machine. The time required (in minutes) to accomplish the task was recorded as follows:

Worker	Domestic Machine	Imported Machine
1	10	9
2	14	12
3	12	12
4	11	15
5	14	10
6	10	11
7	15	12
8	9	8
9	15	10
10	14	15
11	12	14
12	13	11
13	10	13
14	11	12
15	10	12
16	15	10
17	12	10
18	11	13

Would you conclude that the distribution of time required to accomplish the task is the same for both machines? Conduct a Wilcoxon test and use the 0.05 significance level. (Show the appropriate hypotheses and state the decision rule.)

12. A random sample of 12 students showed rankings on writing ability test scores and reasoning ability test scores as follows (1 = lowest score for each test). Compute the correlation of ranks.

Student	*Writing Ability*	*Reasoning Ability*
1	5	4
2	11	11
3	7	1
4	8	10
5	1	2
6	4	5
7	12	12
8	2	3
9	6	8
10	9	6
11	10	9
12	3	7

13. The following information relating to a product was obtained from a random sample of 10 stores. Rank the stores from 1 to 10 according to number of units sold (1 = least number sold) and price (1 = lowest price). Compute the correlation of ranks.

Store	*Number of Units Sold*	*Price (Cents)*
A	162	57
B	193	41
C	150	70
D	173	54
E	184	43
F	155	62
G	171	46
H	180	49
I	176	44
J	164	52

Appendix B

Tables

Table B.1

Table of Areas Under the Normal Curve

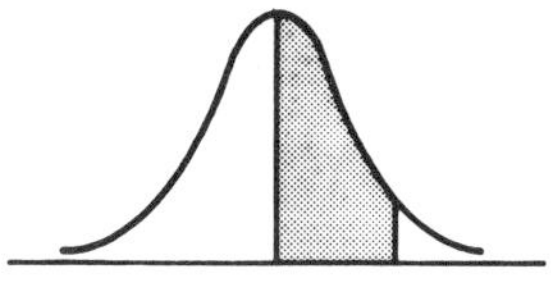

Area from the mean to a specified value of z (the shaded area).

z	**.00**	**.01**	**.02**	**.03**	**.04**	**.05**	**.06**	**.07**	**.08**	**.09**
0.0	.0000	.0040	.0080	.0120	.0160	.0199	.0239	.0279	.0319	.0359
0.1	.0398	.0438	.0478	.0517	.0557	.0596	.0636	.0675	.0714	.0753
0.2	.0793	.0832	.0871	.0910	.0948	.0987	.0026	.1064	.1103	.1141
0.3	.1179	.1217	.1255	.1293	.1331	.1368	.1406	.1443	.1480	.1517
0.4	.1554	.1591	.1628	.1664	.1700	.1736	.1772	.1808	.1844	.1879

Table B.1 (*cont.*)

z	.00	.01	.02	.03	.04	.05	.06	.07	.08	.09
0.5	.1915	.1950	.1985	.2019	.2054	.2088	.2123	.2157	.2190	.2224
0.6	.2257	.2291	.2324	.2357	.2389	.2422	.2454	.2486	.2517	.2549
0.7	.2580	.2611	.2642	.2673	.2704	.2734	.2764	.2794	.2823	.2852
0.8	.2881	.2910	.2939	.2967	.2995	.3023	.3051	.3078	.3106	.3133
0.9	.3159	.3186	.3212	.3238	.3264	.3289	.3315	.3340	.3365	.3389
1.0	.3413	.3438	.3461	.3485	.3508	.3531	.3554	.3577	.3599	.3621
1.1	.3643	.3665	.3686	.3708	.3729	.3749	.3770	.3790	.3810	.3830
1.2	.3849	.3869	.3888	.3907	.3925	.3944	.3962	.3980	.3997	.4015
1.3	.4032	.4049	.4066	.4082	.4099	.4115	.4131	.4147	.4162	.4177
1.4	.4192	.4207	.4222	.4236	.4251	.4265	.4279	.4292	.4306	.4319
1.5	.4332	.4345	.4357	.4370	.4382	.4394	.4406	.4418	.4429	.4441
1.6	.4452	.4463	.4474	.4484	.4495	.4505	.4515	.4525	.4535	.4545
1.7	.4554	.4564	.4573	.4582	.4591	.4599	.4608	.4616	.4625	.4633
1.8	.4641	.4649	.4656	.4664	.4671	.4678	.4686	.4693	.4699	.4706
1.9	.4713	.4719	.4726	.4732	.4738	.4744	.4750	.4756	.4761	.4767
2.0	.4772	.4778	.4783	.4788	.4793	.4798	.4803	.4808	.4812	.4817
2.1	.4821	.4826	.4830	.4834	.4838	.4842	.4846	.4850	.4854	.4857
2.2	.4861	.4864	.4868	.4871	.4875	.4878	.4881	.4884	.4887	.4890
2.3	.4893	.4896	.4898	.4901	.4904	.4906	.4909	.4911	.4913	.4916
2.4	.4918	.4920	.4922	.4925	.4927	.4929	.4931	.4932	.4934	.4936
2.5	.4938	.4940	.4941	.4943	.4945	.4946	.4948	.4949	.4951	.4952
2.6	.4953	.4955	.4956	.4957	.4959	.4960	.4961	.4962	.4963	.4964
2.7	.4965	.4966	.4967	.4968	.4969	.4970	.4971	.4972	.4973	.4974
2.8	.4974	.4975	.4976	.4977	.4977	.4978	.4979	.4979	.4980	.4981
2.9	.4981	.4982	.4982	.4983	.4984	.4984	.4985	.4985	.4986	.4986
3.0	.4987	.4987	.4987	.4988	.4988	.4989	.4989	.4989	.4990	.4990
3.1	.4990	.4991	.4991	.4991	.4992	.4992	.4992	.4992	.4993	.4993
3.2	.4993	.4993	.4994	.4994	.4994	.4994	.4994	.4995	.4995	.4995
3.3	.4995	.4995	.4995	.4996	.4996	.4996	.4996	.4996	.4996	.4997
3.4	.4997	.4997	.4997	.4997	.4997	.4997	.4997	.4997	.4997	.4998

SOURCE: Derived from *Tables of Normal Probability Functions, Applied Mathematics Series 23*, National Bureau of Standards, Washington, D.C.

Table B.2

Chi-square Distribution: Critical Values

Area in the upper tail (the shaded area).

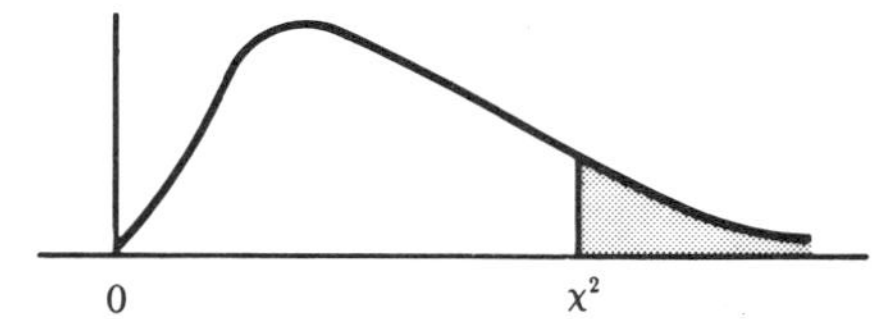

d.f.	Probability of exceeding the critical value							
	.995	.99	.975	.95	.05	.025	.01	.005
1	.0000393	.000157	.000982	.00393	3.841	5.024	6.635	7.879
2	.0100	.0201	.0506	.103	5.991	7.378	9.210	10.597
3	.0717	.115	.216	.352	7.815	9.348	11.345	12.838
4	.207	.297	.484	.711	9.488	11.143	13.277	14.860
5	.412	.554	.831	1.145	11.070	12.832	15.086	16.750
6	.676	.872	1.237	1.635	12.592	14.449	16.812	18.548
7	.989	1.239	1.690	2.167	14.067	16.013	18.475	20.278
8	1.344	1.646	2.180	2.733	15.507	17.535	20.090	21.955
9	1.735	2.088	2.700	3.325	16.919	19.023	21.666	23.589
10	2.156	2.558	3.247	3.940	18.307	20.483	23.209	25.188
11	2.603	3.053	3.816	4.575	19.675	21.920	24.725	26.757
12	3.074	3.571	4.404	5.226	21.026	23.337	26.217	28.300
13	3.565	4.107	5.009	5.892	22.362	24.736	27.688	29.819
14	4.075	4.660	5.629	6.571	23.685	26.119	29.141	31.319
15	4.601	5.229	6.262	7.261	24.996	27.488	30.578	32.801
16	5.142	5.812	6.908	7.962	26.296	28.845	32.000	34.267
17	5.697	6.408	7.564	8.672	27.587	30.191	33.409	35.718
18	6.265	7.015	8.231	9.390	28.869	31.526	34.805	37.156
19	6.844	7.633	8.907	10.117	30.144	32.852	36.191	38.582
20	7.434	8.260	9.591	10.851	31.410	34.170	37.566	39.997
21	8.034	8.897	10.283	11.591	32.671	35.479	38.932	41.401
22	8.643	9.542	10.982	12.338	33.924	36.781	40.289	42.796
23	9.260	10.196	11.689	13.091	35.172	38.076	41.638	44.181
24	9.886	10.856	12.401	13.848	36.415	39.364	42.980	45.558
25	10.520	11.524	13.120	14.611	37.652	40.646	44.314	46.928
26	11.160	12.198	13.844	15.379	38.885	41.923	45.642	48.290
27	11.808	12.879	14.573	16.151	40.113	43.194	46.963	49.645
28	12.461	13.565	15.308	16.928	41.337	44.461	48.278	50.993
29	13.121	14.256	16.047	17.708	42.557	45.722	49.588	52.336
30	13.787	14.953	16.791	18.493	43.773	46.979	50.892	53.672

SOURCE: Table 8 of *Biometrika Tables for Statisticians*, Volume I, by permission of the *Biometrika* trustees.

Table B.3

F Distribution: Critical Values Which Mark Off, in the Upper Tail, 0.05 of the Total Area Under the Distribution Curve

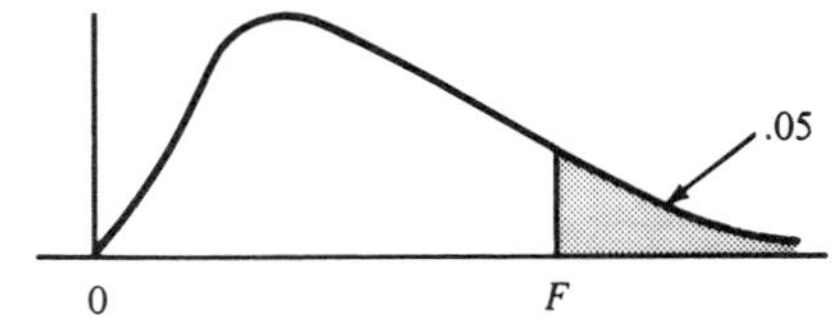

Degrees of freedom for denominator	Degrees of freedom for numerator: 1	2	3	4	5	6	7	8	9	10	12	15	20	24	30	40	60	120	∞
1	161	200	216	225	230	234	237	239	241	242	244	246	248	249	250	251	252	253	254
2	18.5	19.0	19.2	19.2	19.3	19.3	19.4	19.4	19.4	19.4	19.4	19.4	19.4	19.5	19.5	19.5	19.5	19.5	19.5
3	10.1	9.55	9.28	9.12	9.01	8.94	8.89	8.85	8.81	8.79	8.74	8.70	8.66	8.64	8.62	8.59	8.57	8.55	8.53
4	7.71	6.94	6.59	6.39	6.26	6.16	6.09	6.04	6.00	5.96	5.91	5.86	5.80	5.77	5.75	5.72	5.69	5.66	5.63
5	6.61	5.79	5.41	5.19	5.05	4.95	4.88	4.82	4.77	4.74	4.68	4.62	4.56	4.53	4.50	4.46	4.43	4.40	4.37
6	5.99	5.14	4.76	4.53	4.39	4.28	4.21	4.15	4.10	4.06	4.00	3.94	3.87	3.84	3.81	3.77	3.74	3.70	3.67
7	5.59	4.74	4.35	4.12	3.97	3.87	3.79	3.73	3.68	3.64	3.57	3.51	3.44	3.41	3.38	3.34	3.30	3.27	3.23
8	5.32	4.46	4.07	3.84	3.69	3.58	3.50	3.44	3.39	3.35	3.28	3.22	3.15	3.12	3.08	3.04	3.01	2.97	2.93
9	5.12	4.26	3.86	3.63	3.48	3.37	3.29	3.23	3.18	3.14	3.07	3.01	2.94	2.90	2.86	2.83	2.79	2.75	2.71
10	4.96	4.10	3.71	3.48	3.33	3.22	3.14	3.07	3.02	2.98	2.91	2.85	2.77	2.74	2.70	2.66	2.62	2.58	2.54
11	4.84	3.98	3.59	3.36	3.20	3.09	3.01	2.95	2.90	2.85	2.79	2.72	2.65	2.61	2.57	2.53	2.49	2.45	2.40
12	4.75	3.89	3.49	3.26	3.11	3.00	2.91	2.85	2.80	2.75	2.69	2.62	2.54	2.51	2.47	2.43	2.38	2.34	2.30
13	4.67	3.81	3.41	3.18	3.03	2.92	2.83	2.77	2.71	2.67	2.60	2.53	2.46	2.42	2.38	2.34	2.30	2.25	2.21
14	4.60	3.74	3.34	3.11	2.96	2.85	2.76	2.70	2.65	2.60	2.53	2.46	2.39	2.35	2.31	2.27	2.22	2.18	2.13
15	4.54	3.68	3.29	3.06	2.90	2.79	2.71	2.64	2.59	2.54	2.48	2.40	2.33	2.29	2.25	2.20	2.16	2.11	2.07
16	4.49	3.63	3.24	3.01	2.85	2.74	2.66	2.59	2.54	2.49	2.42	2.35	2.28	2.24	2.19	2.15	2.11	2.06	2.01
17	4.45	3.59	3.20	2.96	2.81	2.70	2.61	2.55	2.49	2.45	2.38	2.31	2.23	2.19	2.15	2.10	2.06	2.01	1.96
18	4.41	3.55	3.16	2.93	2.77	2.66	2.58	2.51	2.46	2.41	2.34	2.27	2.19	2.15	2.11	2.06	2.02	1.97	1.92
19	4.38	3.52	3.13	2.90	2.74	2.63	2.54	2.48	2.42	2.38	2.31	2.23	2.16	2.11	2.07	2.03	1.98	1.93	1.88
20	4.35	3.49	3.10	2.87	2.71	2.60	2.51	2.45	2.39	2.35	2.28	2.20	2.12	2.08	2.04	1.99	1.95	1.90	1.84
21	4.32	3.47	3.07	2.84	2.68	2.57	2.49	2.42	2.37	2.32	2.25	2.18	2.10	2.05	2.01	1.96	1.92	1.87	1.81
22	4.30	3.44	3.05	2.82	2.66	2.55	2.46	2.40	2.34	2.30	2.23	2.15	2.07	2.03	1.98	1.94	1.89	1.84	1.78
23	4.28	3.42	3.03	2.80	2.64	2.53	2.44	2.37	2.32	2.27	2.20	2.13	2.05	2.01	1.96	1.91	1.86	1.81	1.76
24	4.26	3.40	3.01	2.78	2.62	2.51	2.42	2.36	2.30	2.25	2.18	2.11	2.03	1.98	1.94	1.89	1.84	1.79	1.73
25	4.24	3.39	2.99	2.76	2.60	2.49	2.40	2.34	2.28	2.24	2.16	2.09	2.01	1.96	1.92	1.87	1.82	1.77	1.71
30	4.17	3.32	2.92	2.69	2.53	2.42	2.33	2.27	2.21	2.16	2.09	2.01	1.93	1.89	1.84	1.79	1.74	1.68	1.62
40	4.08	3.23	2.84	2.61	2.45	2.34	2.25	2.18	2.12	2.08	2.00	1.92	1.84	1.79	1.74	1.69	1.64	1.58	1.51
60	4.00	3.15	2.76	2.53	2.37	2.25	2.17	2.10	2.04	1.99	1.92	1.84	1.75	1.70	1.65	1.59	1.53	1.47	1.39
120	3.92	3.07	2.68	2.45	2.29	2.18	2.09	2.02	1.96	1.91	1.83	1.75	1.66	1.61	1.55	1.50	1.43	1.35	1.25
∞	3.84	3.00	2.60	2.37	2.21	2.10	2.01	1.94	1.88	1.83	1.75	1.67	1.57	1.52	1.46	1.39	1.32	1.22	1.00

SOURCE: M. Merrington and C. M. Thompson, "Tables of percentage points of the inverted beta (F) distribution," *Biometrika*, Vol. 33 (1943), by permission of the *Biometrika* trustees.

Table B.4

F Distribution: Critical Values Which Mark Off, in the Upper Tail, 0.01 of the Total Area Under the Distribution Curve

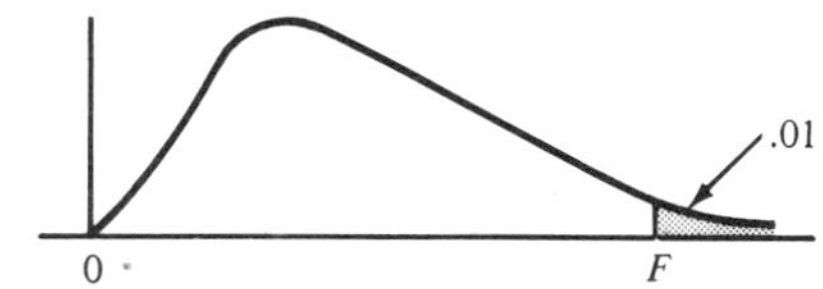

Degrees of freedom for denominator	Degrees of freedom for numerator: 1	2	3	4	5	6	7	8	9	10	12	15	20	24	30	40	60	120	∞
1	4,052	5,000	5,403	5,625	5,764	5,859	5,928	5,982	6,023	6,056	6,106	6,157	6,209	6,235	6,261	6,287	6,313	6,339	6,366
2	98.5	99.0	99.2	99.2	99.3	99.3	99.4	99.4	99.4	99.4	99.4	99.4	99.4	99.5	99.5	99.5	99.5	99.5	99.5
3	34.1	30.8	29.5	28.7	28.2	27.9	27.7	27.5	27.3	27.2	27.1	26.9	26.7	26.6	26.5	26.4	26.3	26.2	26.1
4	21.2	18.0	16.7	16.0	15.5	15.2	15.0	14.8	14.7	14.5	14.4	14.2	14.0	13.9	13.8	13.7	13.7	13.6	13.5
5	16.3	13.3	12.1	11.4	11.0	10.7	10.5	10.3	10.2	10.1	9.89	9.72	9.55	9.47	9.38	9.29	9.20	9.11	9.02
6	13.7	10.9	9.78	9.15	8.75	8.47	8.26	8.10	7.98	7.87	7.72	7.56	7.40	7.31	7.23	7.14	7.06	6.97	6.88
7	12.2	9.55	8.45	7.85	7.46	7.19	6.99	6.84	6.72	6.62	6.47	6.31	6.16	6.07	5.99	5.91	5.82	5.74	5.65
8	11.3	8.65	7.59	7.01	6.63	6.37	6.18	6.03	5.91	5.81	5.67	5.52	5.36	5.28	5.20	5.12	5.03	4.95	4.86
9	10.6	8.02	6.99	6.42	6.06	5.80	5.61	5.47	5.35	5.26	5.11	4.96	4.81	4.73	4.65	4.57	4.48	4.40	4.31
10	10.0	7.56	6.55	5.99	5.64	5.39	5.20	5.06	4.94	4.85	4.71	4.56	4.41	4.33	4.25	4.17	4.08	4.00	3.91
11	9.65	7.21	6.22	5.67	5.32	5.07	4.89	4.74	4.63	4.54	4.40	4.25	4.10	4.02	3.94	3.86	3.78	3.69	3.60
12	9.33	6.93	5.95	5.41	5.06	4.82	4.64	4.50	4.39	4.30	4.16	4.01	3.86	3.78	3.70	3.62	3.54	3.45	3.36
13	9.07	6.70	5.74	5.21	4.86	4.62	4.44	4.30	4.19	4.10	3.96	3.82	3.66	3.59	3.51	3.43	3.34	3.25	3.17
14	8.86	6.51	5.56	5.04	4.70	4.46	4.28	4.14	4.03	3.94	3.80	3.66	3.51	3.43	3.35	3.27	3.18	3.09	3.00
15	8.68	6.36	5.42	4.89	4.56	4.32	4.14	4.00	3.89	3.80	3.67	3.52	3.37	3.29	3.21	3.13	3.05	2.96	2.87
16	8.53	6.23	5.29	4.77	4.44	4.20	4.03	3.89	3.78	3.69	3.55	3.41	3.26	3.18	3.10	3.02	2.93	2.84	2.75
17	8.40	6.11	5.19	4.67	4.34	4.10	3.93	3.79	3.68	3.59	3.46	3.31	3.16	3.08	3.00	2.92	2.83	2.75	2.65
18	8.29	6.01	5.09	4.58	4.25	4.01	3.84	3.71	3.60	3.51	3.37	3.23	3.08	3.00	2.92	2.84	2.75	2.66	2.57
19	8.19	5.93	5.01	4.50	4.17	3.94	3.77	3.63	3.52	3.43	3.30	3.15	3.00	2.92	2.84	2.76	2.67	2.58	2.49
20	8.10	5.85	4.94	4.43	4.10	3.87	3.70	3.56	3.46	3.37	3.23	3.09	2.94	2.86	2.78	2.69	2.61	2.52	2.42
21	8.02	5.78	4.87	4.37	4.04	3.81	3.64	3.51	3.40	3.31	3.17	3.03	2.88	2.80	2.72	2.64	2.55	2.46	2.36
22	7.95	5.72	4.82	4.31	3.99	3.76	3.59	3.45	3.35	3.26	3.12	2.98	2.83	2.75	2.67	2.58	2.50	2.40	2.31
23	7.88	5.66	4.76	4.26	3.94	3.71	3.54	3.41	3.30	3.21	3.07	2.93	2.78	2.70	2.62	2.54	2.45	2.35	2.26
24	7.82	5.61	4.72	4.22	3.90	3.67	3.50	3.36	3.26	3.17	3.03	2.89	2.74	2.66	2.58	2.49	2.40	2.31	2.21
25	7.77	5.57	4.68	4.18	3.86	3.63	3.46	3.32	3.22	3.13	2.99	2.85	2.70	2.62	2.53	2.45	2.36	2.27	2.17
30	7.56	5.39	4.51	4.02	3.70	3.47	3.30	3.17	3.07	2.98	2.84	2.70	2.55	2.47	2.39	2.30	2.21	2.11	2.01
40	7.31	5.18	4.31	3.83	3.51	3.29	3.12	2.99	2.89	2.80	2.66	2.52	2.37	2.29	2.20	2.11	2.02	1.92	1.80
60	7.08	4.98	4.13	3.65	3.34	3.12	2.95	2.82	2.72	2.63	2.50	2.35	2.20	2.12	2.03	1.94	1.84	1.73	1.60
120	6.85	4.79	3.95	3.48	3.17	2.96	2.79	2.66	2.56	2.47	2.34	2.19	2.03	1.95	1.86	1.76	1.66	1.53	1.38
∞	6.63	4.61	3.78	3.32	3.02	2.80	2.64	2.51	2.41	2.32	2.18	2.04	1.88	1.79	1.70	1.59	1.47	1.32	1.00

SOURCE: M. Merrington and C. M. Thompson, "Tables of percentage points of the inverted beta (*F*) distribution," *Biometrika*, Vol. 33 (1943), by permission of the *Biometrika* trustees.

Table B.5

t Distribution: Table of Areas in Both Tails of the Distribution

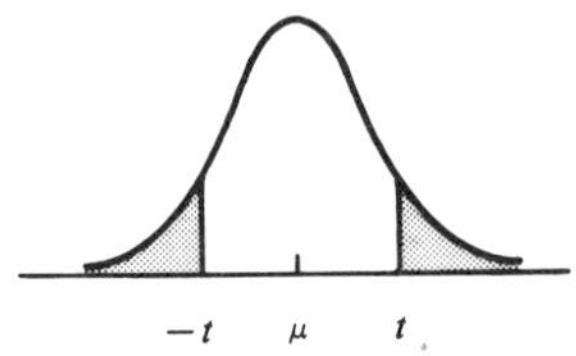

df	Probability (Area in both tails combined)								
	.5	.4	.3	.2	.1	.05	.02	.01	.001
1	1.000	1.376	1.963	3.078	6.314	12.706	31.821	63.657	636.619
2	.816	1.061	1.386	1.886	2.920	4.303	6.965	9.925	31.598
3	.765	.978	1.250	1.638	2.353	3.182	4.541	5.841	12.941
4	.741	.941	1.190	1.533	2.132	2.776	3.747	4.604	8.610
5	.727	.920	1.156	1.476	2.015	2.571	3.365	4.032	6.859
6	.718	.906	1.134	1.440	1.943	2.447	3.143	3.707	5.959
7	.711	.896	1.119	1.415	1.895	2.365	2.998	3.499	5.405
8	.706	.889	1.108	1.397	1.860	2.306	2.896	3.355	5.041
9	.703	.883	1.100	1.383	1.833	2.262	2.821	3.250	4.781
10	.700	.879	1.093	1.372	1.812	2.228	2.764	3.169	4.587
11	.697	.876	1.088	1.363	1.796	2.201	2.718	3.106	4.437
12	.695	.873	1.083	1.356	1.782	2.179	2.681	3.055	4.318
13	.694	.870	1.079	1.350	1.771	2.160	2.650	3.012	4.221
14	.692	.868	1.076	1.345	1.761	2.145	2.624	2.977	4.140
15	.691	.866	1.074	1.341	1.753	2.131	2.602	2.947	4.073
16	.690	.865	1.071	1.337	1.746	2.120	2.583	2.921	4.015
17	.689	.863	1.069	1.333	1.740	2.110	2.567	2.898	3.965
18	.688	.862	1.067	1.330	1.734	2.101	2.552	2.878	3.922
19	.688	.861	1.066	1.328	1.729	2.093	2.539	2.861	3.883
20	.687	.860	1.064	1.325	1.725	2.086	2.528	2.845	3.850
21	.686	.859	1.063	1.323	1.721	2.080	2.518	2.831	3.819
22	.686	.858	1.061	1.321	1.717	2.074	2.508	2.819	3.792
23	.685	.858	1.060	1.319	1.714	2.069	2.500	2.807	3.767
24	.685	.857	1.059	1.318	1.711	2.064	2.492	2.797	3.745
25	.684	.856	1.058	1.316	1.708	2.060	2.485	2.787	3.725
26	.684	.856	1.058	1.315	1.706	2.056	2.479	2.779	3.707
27	.684	.855	1.057	1.314	1.703	2.052	2.473	2.771	3.690
28	.683	.855	1.056	1.313	1.701	2.048	2.467	2.763	3.674
29	.683	.854	1.055	1.311	1.699	2.045	2.462	2.756	3.659
30	.683	.854	1.055	1.310	1.697	2.042	2.457	2.750	3.646
40	.681	.851	1.050	1.303	1.684	2.021	2.423	2.704	3.551
60	.679	.848	1.046	1.296	1.671	2.000	2.390	2.660	3.460
120	.677	.845	1.041	1.289	1.658	1.980	2.358	2.617	3.373
∞	.674	.842	1.036	1.282	1.645	1.960	2.326	2.576	3.291

Table B.5 is taken from Table III of Fisher and Yates: *Statistical Tables for Biological, Agricultural and Medical Research*, published by Longman Group Ltd., London (previously published by Oliver & Boyd, Edinburgh), and by permission of the authors and publishers.

Table B.6

Common Logarithms

N	0	1	2	3	4	5	6	7	8	9
10	0000	0043	0086	0128	0170	0212	0253	0294	0334	0374
11	0414	0453	0492	0531	0569	0607	0645	0682	0719	0755
12	0792	0828	0864	0899	0934	0969	1004	1038	1072	1106
13	1139	1173	1206	1239	1271	1303	1335	1367	1399	1430
14	1461	1492	1523	1553	1584	1614	1644	1673	1703	1732
15	1761	1790	1818	1847	1875	1903	1931	1959	1987	2014
16	2041	2068	2095	2122	2148	2175	2201	2227	2253	2279
17	2304	2330	2355	2380	2405	2430	2455	2480	2504	2529
18	2553	2577	2601	2625	2648	2672	2695	2718	2742	2765
19	2788	2810	2833	2856	2878	2900	2923	2945	2967	2989
20	3010	3032	3054	3075	3096	3118	3139	3160	3181	3201
21	3222	3243	3263	3284	3304	3324	3345	3365	3385	3404
22	3424	3444	3464	3483	3502	3522	3541	3560	3579	3598
23	3617	3636	3655	3674	3692	3711	3729	3747	3766	3784
24	3802	3820	3838	3856	3874	3892	3909	3927	3945	3962
25	3979	3997	4014	4031	4048	4065	4082	4099	4116	4133
26	4150	4166	4183	4200	4216	4232	4249	4265	4281	4298
27	4314	4330	4346	4362	4378	4393	4409	4425	4440	4456
28	4472	4487	4502	4518	4533	4548	4564	4579	4594	4609
29	4624	4639	4654	4669	4683	4698	4713	4728	4742	4757
30	4771	4786	4800	4814	4829	4843	4857	4871	4886	4900
31	4914	4928	4942	4955	4969	4983	4997	5011	5024	5038
32	5051	5065	5079	5092	5105	5119	5132	5145	5159	5172
33	5185	5198	5211	5224	5237	5250	5263	5276	5289	5302
34	5315	5328	5340	5353	5366	5378	5391	5403	5416	5428
35	5441	5453	5465	5478	5490	5502	5514	5527	5539	5551
36	5563	5575	5587	5599	5611	5623	5635	5647	5658	5670
37	5682	5694	5705	5717	5729	5740	5752	5763	5775	5786
38	5798	5809	5821	5832	5843	5855	5866	5877	5888	5899
39	5911	5922	5933	5944	5955	5966	5977	5988	5999	6010
40	6021	6031	6042	6053	6064	6075	6085	6096	6107	6117
41	6128	6138	6149	6160	6170	6180	6191	6201	6212	6222
42	6232	6243	6253	6263	6274	6284	6294	6304	6314	6325
43	6335	6345	6355	6365	6375	6385	6395	6405	6415	6425
44	6435	6444	6454	6464	6474	6484	6493	6503	6513	6522
45	6532	6542	6551	6561	6571	6580	6590	6599	6609	6618
46	6628	6637	6646	6656	6665	6675	6684	6693	6702	6712
47	6721	6730	6739	6749	6758	6767	6776	6785	6794	6803
48	6812	6821	6830	6839	6848	6857	6866	6875	6884	6893
49	6902	6911	6920	6928	6937	6946	6955	6964	6972	6981
50	6990	6998	7007	7016	7024	7033	7042	7050	7059	7067
51	7076	7084	7093	7101	7110	7118	7126	7135	7143	7152
52	7160	7168	7177	7185	7193	7202	7210	7218	7226	7235
53	7243	7251	7259	7267	7275	7284	7292	7300	7308	7316
54	7324	7332	7340	7348	7356	7364	7372	7380	7388	7396

Table B.6 (*cont.*)

N	0	1	2	3	4	5	6	7	8	9
55	7404	7412	7419	7427	7435	7443	7451	7459	7466	7474
56	7482	7490	7497	7505	7513	7520	7528	7536	7543	7551
57	7559	7566	7574	7582	7589	7597	7604	7612	7619	7627
58	7634	7642	7649	7657	7664	7672	7679	7686	7694	7701
59	7709	7716	7723	7731	7738	7745	7752	7760	7767	7774
60	7782	7789	7796	7803	7810	7818	7825	7832	7839	7846
61	7853	7860	7868	7875	7882	7889	7896	7903	7910	7917
62	7924	7931	7938	7945	7952	7959	7966	7973	7980	7987
63	7993	8000	8007	8014	8021	8028	8035	8041	8048	8055
64	8062	8069	8075	8082	8089	8096	8102	8109	8116	8122
65	8129	8136	8142	8149	8156	8162	8169	8176	8182	8189
66	8195	8202	8209	8215	8222	8228	8235	8241	8248	8254
67	8261	8267	8274	8280	8287	8293	8299	8306	8312	8319
68	8325	8331	8338	8344	8351	8357	8363	8370	8376	8382
69	8388	8395	8401	8407	8414	8420	8426	8432	8439	8445
70	8451	8457	8463	8470	8476	8482	8488	8494	8500	8506
71	8513	8519	8525	8531	8537	8543	8549	8555	8561	8567
72	8573	8579	8585	8591	8597	8603	8609	8615	8621	8627
73	8633	8639	8645	8651	8657	8663	8669	8675	8681	8686
74	8692	8698	8704	8710	8716	8722	8727	8733	8739	8745
75	8751	8756	8762	8768	8774	8779	8785	8791	8797	8802
76	8808	8814	8820	8825	8831	8837	8842	8848	8854	8859
77	8865	8871	8876	8882	8887	8893	8899	8904	8910	8915
78	8921	8927	8932	8938	8943	8949	8954	8960	8965	8971
79	8976	8982	8987	8993	8998	9004	9009	9015	9020	9025
80	9031	9036	9042	9047	9053	9058	9063	9069	9074	9079
81	9085	9090	9096	9101	9106	9112	9117	9122	9128	9133
82	9138	9143	9149	9154	9159	9165	9170	9175	9180	9186
83	9191	9196	9201	9206	9212	9217	9222	9227	9232	9238
84	9243	9248	9253	9258	9263	9269	9274	9279	9284	9289
85	9294	9299	9304	9309	9315	9320	9325	9330	9335	9340
86	9345	9350	9355	9360	9365	9370	9375	9380	9385	9390
87	9395	9400	9405	9410	9415	9420	9425	9430	9435	9440
88	9445	9450	9455	9460	9465	9469	9474	9479	9484	9489
89	9494	9499	9504	9509	9513	9518	9523	9528	9533	9538
90	9542	9547	9552	9557	9562	9566	9571	9576	9581	9586
91	9590	9595	9600	9605	9609	9614	9619	9624	9628	9633
92	9638	9643	9647	9652	9657	9661	9666	9671	9675	9680
93	9685	9689	9694	9699	9703	9708	9713	9717	9722	9727
94	9731	9736	9741	9745	9750	9754	9759	9763	9768	9773
95	9777	9782	9786	9791	9795	9800	9805	9809	9814	9818
96	9823	9827	9832	9836	9841	9845	9850	9854	9859	9863
97	9868	9872	9877	9881	9886	9890	9894	9899	9903	9908
98	9912	9917	9921	9926	9930	9934	9939	9943	9948	9952
99	9956	9961	9965	9969	9974	9978	9983	9987	9991	9996

Table B.7

Table of Squares

N	.0	.1	.2	.3	.4	.5	.6	.7	.8	.9
0	0.00	0.01	0.04	0.09	0.16	0.25	0.36	0.49	0.64	0.81
1	1.00	1.21	1.44	1.69	1.96	2.25	2.56	2.89	3.24	3.61
2	4.00	4.41	4.84	5.29	5.76	6.25	6.76	7.29	7.84	8.41
3	9.00	9.61	10.24	10.89	11.56	12.25	12.96	13.69	14.44	15.21
4	16.00	16.81	17.64	18.49	19.36	20.25	21.16	22.09	23.04	24.01
5	25.00	26.01	27.04	28.09	29.16	30.25	31.36	32.49	33.64	34.81
6	36.00	37.21	38.44	39.69	40.96	42.25	43.56	44.89	46.24	47.61
7	49.00	50.41	51.84	53.29	54.76	56.25	57.76	59.29	60.84	62.41
8	64.00	65.61	67.24	68.89	70.56	72.25	73.96	75.69	77.44	79.21
9	81.00	82.81	84.64	86.49	88.36	90.25	92.16	94.09	96.04	98.01
10	100.00	102.01	104.04	106.09	108.16	110.25	112.36	114.49	116.64	118.81
11	121.00	123.21	125.44	127.69	129.96	132.25	134.56	136.89	139.24	141.61
12	144.00	146.41	148.84	151.29	153.76	156.25	158.76	161.29	163.84	166.41
13	169.00	171.61	174.24	176.89	179.56	182.25	184.96	187.69	190.44	193.21
14	196.00	198.81	201.64	204.49	207.36	210.25	213.16	216.09	219.04	222.01
15	225.00	228.01	231.04	234.09	237.16	240.25	243.36	246.49	249.64	252.81
16	256.00	259.21	262.44	265.69	268.96	272.25	275.56	278.89	282.24	285.61
17	289.00	292.41	295.84	299.29	302.76	306.25	309.76	313.29	316.84	320.41
18	324.00	327.61	331.24	334.89	338.56	342.25	345.96	349.69	353.44	357.21
19	361.00	364.81	368.64	372.49	376.36	380.25	384.16	388.09	392.04	396.01
20	400.00	404.01	408.04	412.09	416.16	420.25	424.36	428.49	432.64	436.81
21	441.00	445.21	449.44	453.69	457.96	462.25	466.56	470.89	475.24	479.61
22	484.00	488.41	492.84	497.29	501.76	506.25	510.76	515.29	519.84	524.41
23	529.00	533.61	538.24	542.89	547.56	552.25	556.96	561.69	566.44	571.21

Table B.7 (*cont.*)

24	576.00	580.81	585.64	590.49	595.36	600.25	605.16	610.09	615.04	620.01
25	625.00	630.01	635.04	640.09	645.16	650.25	655.36	660.49	665.64	670.81
26	676.00	681.21	686.44	691.69	696.96	702.25	707.56	712.89	718.24	723.61
27	729.00	734.41	739.84	745.29	750.76	756.25	761.76	767.29	772.84	778.41
28	784.00	789.61	795.24	800.89	806.56	812.25	817.96	823.69	829.44	835.21
29	841.00	846.81	852.64	858.49	864.36	870.25	876.16	882.09	888.04	894.01
30	900.00	906.01	912.04	918.09	924.16	930.25	936.36	942.49	948.64	954.81
31	961.00	967.21	973.44	979.69	985.96	992.25	998.56	1004.89	1011.24	1017.61
32	1024.00	1030.41	1036.84	1043.29	1049.76	1056.25	1062.76	1069.29	1075.84	1082.41
33	1089.00	1095.61	1102.24	1108.89	1115.56	1122.25	1128.96	1135.69	1142.44	1149.21
34	1156.00	1162.81	1169.64	1176.49	1183.36	1190.25	1197.16	1204.09	1211.04	1218.01
35	1225.00	1232.01	1239.04	1246.09	1253.16	1260.25	1267.36	1274.49	1281.64	1288.81
36	1296.00	1303.21	1310.44	1317.69	1324.96	1332.25	1339.56	1346.89	1354.24	1361.61
37	1369.00	1376.41	1383.84	1391.29	1398.76	1406.25	1413.76	1421.29	1428.84	1436.41
38	1444.00	1451.61	1459.24	1466.89	1474.56	1482.25	1489.96	1497.69	1505.44	1513.21
39	1521.00	1528.81	1536.64	1544.49	1552.36	1560.25	1568.16	1576.09	1584.04	1592.01
40	1600.00	1608.01	1616.04	1624.09	1632.16	1640.25	1648.36	1656.49	1664.64	1672.81
41	1681.00	1689.21	1697.44	1705.69	1713.96	1722.25	1730.56	1738.89	1747.24	1755.61
42	1764.00	1772.41	1780.84	1789.29	1797.76	1806.25	1814.76	1823.29	1831.84	1840.41
43	1849.00	1857.61	1866.24	1874.89	1883.56	1892.25	1900.96	1909.69	1918.44	1927.21
44	1936.00	1944.81	1953.64	1962.49	1971.36	1980.25	1989.16	1998.09	2007.04	2016.01
45	2025.00	2034.01	2043.04	2052.09	2061.16	2070.25	2079.36	2088.49	2097.64	2106.81
46	2116.00	2125.21	2134.44	2143.69	2152.96	2162.25	2171.56	2180.89	2190.24	2199.61
47	2209.00	2218.41	2227.84	2237.29	2246.76	2256.25	2265.76	2275.29	2284.84	2294.41
48	2304.00	2313.61	2323.24	2332.89	2342.56	2352.25	2361.96	2371.69	2381.44	2391.21
49	2401.00	2410.81	2420.64	2430.49	2440.36	2450.25	2460.16	2470.09	2480.04	2490.01
50	2500.00	2510.01	2520.04	2530.09	2540.16	2550.25	2560.36	2570.49	2580.64	2590.81
51	2601.00	2611.21	2621.44	2631.69	2641.96	2652.25	2662.56	2672.89	2683.24	2693.61
52	2704.00	2714.41	2724.84	2735.29	2745.76	2756.25	2766.76	2777.29	2787.84	2798.41
53	2809.00	2819.61	2830.24	2840.89	2851.56	2862.25	2872.96	2883.69	2894.44	2905.21
54	2916.00	2926.81	2937.64	2948.49	2959.36	2970.25	2981.16	2992.09	3003.04	3014.01

Table B.7 (*cont.*)

Table of Squares

N	.0	.1	.2	.3	.4	.5	.6	.7	.8	.9
55	3025.00	3036.01	3047.04	3058.09	3069.16	3080.25	3091.36	3102.49	3113.64	3124.81
56	3136.00	3147.21	3158.44	3169.69	3180.96	3192.25	3203.56	3214.89	3226.24	3237.61
57	3249.00	3260.41	3271.84	3283.29	3294.76	3306.25	3317.76	3329.29	3340.84	3352.41
58	3364.00	3375.61	3387.24	3398.89	3410.56	3422.25	3433.96	3445.69	3457.44	3469.21
59	3481.00	3492.81	3504.64	3516.49	3528.36	3540.25	3552.16	3564.09	3576.04	3588.01
60	3600.00	3612.01	3624.04	3636.09	3648.16	3660.25	3672.36	3684.49	3696.64	3708.81
61	3721.00	3733.21	3745.44	3757.69	3769.96	3782.25	3794.56	3806.89	3819.24	3831.61
62	3844.00	3856.41	3868.84	3881.29	3893.76	3906.25	3918.76	3931.29	3943.84	3956.41
63	3969.00	3981.61	3994.24	4006.89	4019.56	4032.25	4044.96	4057.69	4070.44	4083.21
64	4096.00	4108.81	4121.64	4134.49	4147.36	4160.25	4173.16	4186.09	4199.04	4212.01
65	4225.00	4238.01	4251.04	4264.09	4277.16	4290.25	4303.36	4316.49	4329.64	4342.81
66	4356.00	4369.21	4382.44	4395.69	4408.96	4422.25	4435.56	4448.89	4462.24	4475.61
67	4489.00	4502.41	4515.84	4529.29	4542.76	4556.25	4569.76	4583.29	4596.84	4610.41
68	4624.00	4637.61	4651.24	4664.89	4678.56	4692.25	4705.96	4719.69	4733.44	4747.21
69	4761.00	4774.81	4788.64	4802.49	4816.36	4830.25	4844.16	4858.09	4872.04	4886.01
70	4900.00	4914.01	4928.04	4942.09	4956.16	4970.25	4984.36	4998.49	5012.64	5026.81
71	5041.00	5055.21	5069.44	5083.69	5097.96	5112.25	5126.56	5140.89	5155.24	5169.61
72	5184.00	5198.41	5212.84	5227.29	5241.76	5256.25	5270.76	5285.29	5299.84	5314.41
73	5329.00	5343.61	5358.24	5372.89	5387.56	5402.25	5416.96	5431.69	5446.44	5461.21
74	5476.00	5490.81	5505.64	5520.49	5535.36	5550.25	5565.16	5580.09	5595.04	5610.01
75	5625.00	5640.01	5655.04	5670.09	5685.16	5700.25	5715.36	5730.49	5745.64	5760.81
76	5776.00	5791.21	5806.44	5821.69	5836.96	5852.25	5867.56	5882.89	5898.24	5913.61
77	5929.00	5944.41	5959.84	5975.29	5990.76	6006.25	6021.76	6037.29	6052.84	6068.41
78	6084.00	6099.61	6115.24	6130.89	6146.56	6162.25	6177.96	6193.69	6209.44	6225.21

Table B.7 (*cont.*)

79	6241.00	6256.81	6272.64	6288.49	6304.36	6320.25	6336.16	6352.09	6368.04	6384.01
80	6400.00	6416.01	6432.04	6448.09	6464.16	6480.25	6496.36	6512.49	6528.64	6544.81
81	6561.00	6577.21	6593.44	6609.69	6625.96	6642.25	6658.56	6674.89	6691.24	6707.61
82	6724.00	6740.41	6756.84	6773.29	6789.76	6806.25	6822.76	6839.29	6855.84	6872.41
83	6889.00	6905.61	6922.24	6938.89	6955.56	6972.25	6988.96	7005.69	7022.44	7039.21
84	7056.00	7072.81	7089.64	7106.49	7123.36	7140.25	7157.16	7174.09	7191.04	7208.01
85	7225.00	7242.01	7259.04	7276.09	7293.16	7310.25	7327.36	7344.49	7361.64	7378.81
86	7396.00	7413.21	7430.44	7447.69	7464.96	7482.25	7499.56	7516.89	7534.24	7551.61
87	7569.00	7586.41	7603.84	7621.29	7638.76	7656.25	7673.76	7691.29	7708.84	7726.41
88	7744.00	7761.61	7779.24	7796.89	7814.56	7832.25	7849.96	7867.69	7885.44	7903.21
89	7921.00	7938.81	7956.64	7974.49	7992.36	8010.25	8028.16	8046.09	8064.04	8082.01
90	8100.00	8118.01	8136.04	8154.09	8172.16	8190.25	8208.36	8226.49	8244.64	8262.81
91	8281.00	8299.21	8317.44	8335.69	8353.96	8372.25	8390.56	8408.89	8427.24	8445.61
92	8464.00	8482.41	8500.84	8519.29	8537.76	8556.25	8574.76	8593.29	8611.84	8630.41
93	8649.00	8667.61	8686.24	8704.89	8723.56	8742.25	8760.96	8779.69	8798.44	8817.21
94	8836.00	8854.81	8873.64	8892.49	8911.36	8930.25	8949.16	8968.09	8987.04	9006.01
95	9025.00	9044.01	9063.04	9082.09	9101.16	9120.25	9139.36	9158.49	9177.64	9196.81
96	9216.00	9235.21	9254.44	9273.69	9292.96	9312.25	9331.56	9350.89	9370.24	9389.61
97	9409.00	9428.41	9447.84	9467.29	9486.76	9506.25	9525.76	9545.29	9564.84	9584.41
98	9604.00	9623.61	9643.24	9662.89	9682.56	9702.25	9721.96	9741.69	9761.44	9781.21
99	9801.00	9820.81	9840.64	9860.49	9880.36	9900.25	9920.16	9940.09	9960.04	9980.01

Table B.8

Table of Square Roots

N	.0	.1	.2	.3	.4	.5	.6	.7	.8	.9
0	0.0000	0.3162	0.4472	0.5477	0.6325	0.7071	0.7746	0.8367	0.8944	0.9487
1	1.0000	1.0488	1.0954	1.1402	1.1832	1.2247	1.2649	1.3038	1.3416	1.3784
2	1.4142	1.4491	1.4832	1.5166	1.5492	1.5811	1.6125	1.6432	1.6733	1.7029
3	1.7320	1.7607	1.7889	1.8166	1.8439	1.8708	1.8974	1.9235	1.9494	1.9748
4	2.0000	2.0248	2.0494	2.0736	2.0976	2.1213	2.1448	2.1679	2.1909	2.2136
5	2.2361	2.2583	2.2803	2.3022	2.3238	2.3452	2.3664	2.3875	2.4083	2.4290
6	2.4495	2.4698	2.4900	2.5100	2.5298	2.5495	2.5690	2.5884	2.6077	2.6268
7	2.6458	2.6646	2.6833	2.7018	2.7203	2.7386	2.7568	2.7749	2.7928	2.8107
8	2.8284	2.8460	2.8636	2.8810	2.8983	2.9155	2.9326	2.9496	2.9665	2.9833
9	3.0000	3.0166	3.0331	3.0496	3.0659	3.0822	3.0984	3.1145	3.1305	3.1464
10	3.1623	3.1780	3.1937	3.2094	3.2249	3.2404	3.2558	3.2711	3.2863	3.3015
11	3.3166	3.3317	3.3466	3.3615	3.3764	3.3912	3.4059	3.4205	3.4351	3.4496
12	3.4641	3.4785	3.4928	3.5071	3.5214	3.5355	3.5496	3.5637	3.5777	3.5917
13	3.6055	3.6194	3.6332	3.6469	3.6606	3.6742	3.6878	3.7014	3.7148	3.7283
14	3.7417	3.7550	3.7683	3.7815	3.7947	3.8079	3.8210	3.8341	3.8471	3.8601
15	3.8730	3.8859	3.8987	3.9115	3.9243	3.9370	3.9497	3.9623	3.9749	3.9875
16	4.0000	4.0125	4.0249	4.0373	4.0497	4.0620	4.0743	4.0866	4.0988	4.1110
17	4.1231	4.1352	4.1473	4.1593	4.1713	4.1833	4.1952	4.2071	4.2190	4.2308
18	4.2426	4.2544	4.2661	4.2778	4.2895	4.3012	4.3128	4.3243	4.3359	4.3474
19	4.3589	4.3704	4.3818	4.3932	4.4045	4.4159	4.4272	4.4385	4.4497	4.4609
20	4.4721	4.4833	4.4944	4.5056	4.5166	4.5277	4.5387	4.5497	4.5607	4.5716
21	4.5826	4.5935	4.6043	4.6152	4.6260	4.6368	4.6476	4.6583	4.6690	4.6797
22	4.6904	4.7011	4.7117	4.7223	4.7329	4.7434	4.7539	4.7644	4.7749	4.7854
23	4.7958	4.8062	4.8166	4.8270	4.8374	4.8477	4.8580	4.8683	4.8785	4.8888

Table B.8 (*cont.*)

24	4.8990	4.9092	4.9193	4.9295	4.9396	4.9497	4.9598	4.9699	4.9800	4.9900
25	5.0000	5.0100	5.0200	5.0299	5.0398	5.0498	5.0596	5.0695	5.0794	5.0892
26	5.0990	5.1088	5.1186	5.1284	5.1381	5.1478	5.1575	5.1672	5.1769	5.1865
27	5.1962	5.2058	5.2154	5.2249	5.2345	4.2440	5.2536	5.2631	5.2726	5.2820
28	5.2915	5.3009	5.3104	5.3198	5.3292	5.3385	5.3479	5.3572	5.3666	5.3759
29	5.3852	5.3944	5.4037	5.4129	5.4222	5.4314	5.4406	5.4498	5.4589	5.4681
30	5.4772	5.4863	5.4955	5.5045	5.5136	5.5227	5.5317	5.5408	5.5498	5.5588
31	5.5678	5.5767	5.5857	5.5946	5.6036	5.6125	5.6214	5.6303	5.6391	5.6480
32	5.6569	5.6657	5.6745	5.6833	5.6921	5.7009	5.7096	5.7184	5.7271	5.7359
33	5.7446	5.7533	5.7619	5.7706	5.7793	5.7879	5.7965	5.8052	5.8138	5.8224
34	5.8310	5.8395	5.8481	5.8566	5.8651	5.8737	5.8822	5.8907	5.8991	5.9076
35	5.9161	5.9245	5.9330	5.9414	5.9498	5.9582	5.9666	5.9749	5.9833	5.9917
36	6.0000	6.0083	6.0166	6.0249	6.0332	6.0415	6.0498	6.0581	6.0663	6.0745
37	6.0828	6.0910	6.0992	6.1074	6.1156	6.1237	6.1319	6.1400	6.1482	6.1563
38	6.1644	6.1725	6.1806	6.1887	6.1968	6.2048	6.2129	6.2209	6.2290	6.2370
39	6.2450	6.2530	6.2610	6.2690	6.2769	6.2849	6.2929	6.3008	6.3087	6.3166
40	6.3246	6.3325	6.3403	6.3482	6.3561	6.3640	6.3718	6.3797	6.3875	6.3953
41	6.4031	6.4109	6.4187	6.4265	6.4343	6.4420	6.4498	6.4576	6.4653	6.4730
42	6.4807	6.4884	6.4962	6.5038	6.5115	6.5192	6.5269	6.5345	6.5422	6.5498
43	6.5574	6.5651	6.5727	6.5803	6.5879	6.5955	6.6030	6.6106	6.6182	6.6257
44	6.6332	6.6408	6.6483	6.6558	6.6633	6.6708	6.6783	6.6858	6.6933	6.7007
45	6.7082	6.7157	6.7231	6.7305	6.7380	6.7454	6.7528	6.7602	6.7676	6.7750
46	6.7823	6.7897	6.7971	6.8044	6.8118	6.8191	6.8264	6.8337	6.8411	6.8484
47	6.8557	6.8629	6.8702	6.8775	6.8848	6.8920	6.8993	6.9065	6.9138	6.9210
48	6.9282	6.9354	6.9426	6.9498	6.9570	6.9642	6.9714	6.9785	6.9857	6.9929
49	7.0000	7.0071	7.0143	7.0214	7.0285	7.0356	7.0427	7.0498	7.0569	7.0640
50	7.0711	7.0781	7.0852	7.0922	7.0993	7.1063	7.1134	7.1204	7.1274	7.1344
51	7.1414	7.1484	7.1554	7.1624	7.1694	7.1763	7.1833	7.1903	7.1972	6.2042
52	7.2111	7.2180	7.2250	7.2319	7.2388	7.2457	7.2526	7.2595	7.2664	7.2732
53	7.2801	7.2870	7.2938	7.3007	7.3075	7.3144	7.3212	7.3280	7.3348	7.3417
54	7.3485	7.3553	7.3621	7.3689	7.3756	7.3824	7.3892	7.3959	7.4027	7.4095

Table B.8 (*cont.*)

Table of Square Roots

N	.0	.1	.2	.3	.4	.5	.6	.7	.8	.9
55	7.4162	7.4229	7.4297	7.4364	7.4431	7.4498	7.4565	7.4632	7.4699	7.4766
56	7.4833	7.4900	7.4967	7.5033	7.5100	7.5166	7.5233	7.5299	7.5366	7.5432
57	7.5498	7.5565	7.5631	7.5697	7.5763	7.5829	7.5895	7.5960	7.6026	7.6092
58	7.6158	7.6223	7.6289	7.6354	7.6420	7.6485	7.6551	7.6616	7.6681	7.6746
59	7.6811	7.6877	7.6942	7.7006	7.7071	7.7136	7.7201	7.7266	7.7330	7.7395
60	7.7460	7.7524	7.7589	7.7653	7.7717	7.7782	7.7846	7.7910	7.7974	7.8038
61	7.8102	7.8166	7.8230	7.8294	7.8358	7.8422	7.8486	7.8549	7.8613	7.8677
62	7.8740	7.8804	7.8867	7.8930	7.8994	7.9057	7.9120	7.9183	7.9246	7.9309
63	7.9373	7.9435	7.9498	7.9561	7.9624	7.9687	7.9750	7.9812	7.9875	7.9937
64	8.0000	8.0062	8.0125	8.0187	8.0250	8.0312	8.0374	8.0436	8.0498	8.0561
65	8.0623	8.0685	8.0746	8.0808	8.0870	8.0932	8.0994	8.1056	8.1117	8.1179
66	8.1240	8.1302	8.1363	8.1425	8.1486	8.1548	8.1609	8.1670	8.1731	8.1792
67	8.1854	8.1915	8.1976	8.2037	8.2097	8.2158	8.2219	8.2280	8.2341	8.2401
68	8.2462	8.2523	8.2583	8.2644	8.2704	8.2765	8.2825	8.2885	8.2946	8.3006
69	8.3066	8.3126	8.3187	8.3247	8.3307	8.3367	8.3427	8.3486	8.3546	8.3606
70	8.3666	8.3726	8.3785	8.3845	8.3905	8.3964	8.4024	8.4083	8.4143	8.4202
71	8.4261	8.4321	8.4380	8.4439	8.4498	8.4558	8.4617	8.4676	8.4735	8.4794
72	8.4853	8.4912	8.4971	8.5029	8.5088	8.5147	8.5206	8.5264	8.5323	8.5381
73	8.5440	8.5499	8.5557	8.5615	8.5674	8.5732	8.5790	8.5849	8.5907	8.5965
74	8.6023	8.6081	8.6139	8.6197	8.6255	8.6313	8.6371	8.6429	8.6487	8.6545
75	8.6603	8.6660	8.6718	8.6776	8.6833	8.6891	8.6948	8.7006	8.7063	8.7121
76	8.7178	8.7235	8.7293	8.7350	8.7407	8.7464	8.7521	8.7579	8.7636	8.7693
77	8.7750	8.7807	8.7864	8.7920	8.7977	8.8034	8.8091	8.8148	8.8204	8.8261
78	8.8318	8.8374	8.8431	8.8487	8.8544	8.8600	8.8657	8.8713	8.8769	8.8826

Table B.8 (*cont.*)

79	8.8882	8.8938	8.8994	8.9051	8.9107	8.9163	8.9219	8.9275	8.9331	8.9387
80	8.9443	8.9499	8.9554	8.9610	8.9666	8.9722	8.9777	8.9833	8.9889	8.9944
81	9.0000	9.0056	9.0111	9.0166	9.0222	9.0277	9.0333	9.0388	9.0443	9.0499
82	9.0554	9.0609	9.0664	9.0719	9.0774	9.0829	9.0885	9.0940	9.0994	9.1049
83	9.1104	9.1159	9.1214	9.1269	9.1324	9.1378	9.1433	9.1488	9.1542	9.1597
84	9.1651	9.1706	9.1761	9.1815	9.1869	9.1924	9.1978	9.2033	9.2087	9.2141
85	9.2195	9.2250	9.2304	9.2358	9.2412	9.2466	9.2520	9.2574	9.2628	9.2682
86	9.2736	9.2790	9.2844	9.2898	9.2952	9.3005	9.3059	9.3113	9.3166	9.3220
87	9.3274	9.3327	9.3381	9.3434	9.3488	9.3541	9.3595	9.3648	9.3702	9.3755
88	9.3808	9.3862	9.3915	9.3968	9.4021	9.4074	9.4128	9.4181	9.4234	9.4287
89	9.4340	9.4393	9.4446	9.4499	9.4552	9.4604	9.4657	9.4710	9.4763	9.4816
90	9.4868	9.4921	9.4974	9.5026	9.5079	9.5131	9.5184	9.5237	9.5289	9.5341
91	9.5394	9.5446	9.5499	9.5551	9.5603	9.5656	9.5708	9.5760	9.5812	9.5864
92	9.5917	9.5969	9.6021	9.6073	9.6125	9.6177	9.6229	9.6281	9.6333	9.6385
93	9.6436	9.6488	9.6540	9.6592	9.6644	9.6695	9.6747	9.6799	9.6850	9.6902
94	9.6954	9.7005	9.7057	9.7108	9.7160	9.7211	9.7263	9.7314	9.7365	9.7417
95	9.7468	9.7519	9.7570	9.7622	9.7673	9.7724	9.7775	9.7826	9.7877	9.7929
96	9.7980	9.8031	9.8082	9.8133	9.8183	9.8234	9.8285	9.8336	9.8387	9.8438
97	9.8489	9.8539	9.8590	9.8641	9.8691	9.8742	9.8793	9.8843	9.8894	9.8944
98	9.8995	9.9045	9.9096	9.9146	9.9197	9.9247	9.9298	9.9348	9.9398	9.9448
99	9.9499	9.9549	9.9599	9.9649	9.9700	9.9750	9.9800	9.9850	9.9900	9.9950

Appendix C

Answers to Even-Numbered Problems

Chapter 3

2. (a) A population of balances due of delinquent accounts
(b) A population of major intersections in the city
(c) A population of IQ scores for the seniors in the high school
A sample of 25 IQ scores
(d) A population of reaction times for adults in the 40 to 50 years age group, a subpopulation of male reaction times and a subpopulation of female reaction times in this age bracket; A sample of 60 adult reaction times, a sample of 30 male reaction times, and a sample of 30 female reaction times

6. Inductive statistics

8. He is estimating population characteristics based on sample data.

Chapter 4

4. Nominal: equivalence
Ordinal: equivalence, relative importance
Interval: equivalence, relative importance, measure of how much more
Ratio: equivalence, relative importance, measure of how much more, true zero point.

6. Interval scale

10. (a) Discrete, because it is measured to the nearest tenth of a percent and so the scale of measurement contains gaps
(b) Ratio scale, because it contains a true zero point

Chapter 5

4. No. Ranking is of limited aid in understanding the distribution of as large a number as 300 quantities.
6. A frequency distribution because it would summarize as well as organize the data
8. The low scores of 62 and 63 do not appear to be as low as they are and the high scores of 97–99 do not appear to be as high as they are, since the scores with zero frequency are omitted.
10. Yes. You do not know the actual 50 measurements. You know only the classes they fall in.
12. (a) 18 (b) 25, 17, 0
14. (a) 14, 37, 57, 60, 29, 17
16. No, the classes overlap. Nonoverlapping classes should be used such as 40–44, 45–49, etc.
18. 0, 2, 10, 22, 25; 78%
20. Cumulative number: 9, 24, 78, 127, 188, 200
Cumulative percentage: 4.5, 12.0, 39.0, 63.5, 94.0, 100.0

Chapter 6

4. Decrease, because the cumulative frequencies become smaller and smaller for higher and higher distances (as shown in Table 5.12)

Chapter 7

2. To describe the distribution of heights by determining such characteristics as the average height, how the heights vary, the proportion of heights which may be considered short, etc.
8. Colors arranged in order of the numbers of shoes sold. This would provide a chart with the most meaningful display of the data.

Chapter 8

2. (a) False (b) False (c) False (d) True (e) False (f) True
(g) False (h) True (i) True
4. Mean and median
6. Mean = 8, median and mode = 5
8. $200
10. (a) 15 oz (b) 13 oz
12. 480 lb
16. Yes, since the same amount ($500) was invested at each percentage
20. (a) 82% (b) 75% (c) 30% (d) 50% (e) 80% (f) 75%
26. (a) Yes (b) No (c) Only if it changes the weight which appears most often
28. (a) Against (b) Against
30. 11 micrometers
32. (a) 5 lb (b) 8 lb (c) 3 lb

Chapter 9

2. No, the score variation could differ.
4. (a) 7 (b) 6, 4–10 (c) $-3, -1, -1, 2, 3$
6. (a) 0 (b) 0

8. Mean = \$110.15
Standard deviation = \$16.20
10. (a) Mean = 30 ml, standard deviation unchanged
(b) Mean = 27.5 ml, standard deviation = 5.5 ml
14. Soybean oil prices (relative variation = 16.7% compared with 25.0% for cottonseed prices)
16. Test scores show less relative variation (30.0% compared with 33.3% for reaction times)
18. Years of experience show more relative variation than ages (42.9% compared to 9.1%)
20. (a) −19.2, 1.8
(b) 4.05 micrometers is unusual since it is 19.2 standard deviations below the mean for the batch.
22. (a) Chemistry 1.00, French 1.50, History 2.00
(b) History, 2.00 standard deviations above the mean
24. (a) 50 to 70 accidents (b) 30 to 90 accidents (c) 15 to 95 accidents
26. 95%

Chapter 10

2. Yes, only one can occur.
6. (a) 0.10 (b) 0.90
8. (a) 0.20 (b) 0.40 (c) 0.80 (d) 0.30 (e) 0.70
12. (a) 1/20 (b) 1/20
18. (a) 1 to 9 (b) 7 to 3

Chapter 12

4. (a) 6
(b)

Plant Counts in Sample	*Sample Mean*
6,2	4.0
6,9	7.5
6,10	8.0
6,3	4.5
2,9	5.5
2,10	6.0
2,3	2.5
9,10	9.5
9,3	6.0
10,3	6.5

(c)

Sample Mean	*Frequency*	*Proportion*
2.5	1	$\frac{1}{10}$
4.0	1	$\frac{1}{10}$
4.5	1	$\frac{1}{10}$
5.5	1	$\frac{1}{10}$

(c)

Sample Mean	Frequency	Proportion
6.0	2	$\frac{2}{10}$
6.5	1	$\frac{1}{10}$
7.5	1	$\frac{1}{10}$
8.0	1	$\frac{1}{10}$
9.5	1	$\frac{1}{10}$
Total	10	

(d) 6/10 (e) 4/10

6. (a) 4.6

(b)

Sample Mean	Probability
2.5	0.10
3.5	0.10
4.0	0.20
4.5	0.20
5.0	0.10
5.5	0.10
6.0	0.10
6.5	0.10
	1.00

(c)

Sample Mean	Probability
3.33	0.10
3.67	0.10
4.00	0.10
4.33	0.10
4.67	0.20
5.00	0.20
5.33	0.10
6.00	0.10
	1.00

(d) Sample of 2 cases: 0.70; Sample of 3 cases: 0.80
(e) Sample of 2 cases: 0.40; Sample of 3 cases: 0.50

Chapter 13

2. (a) 160–240 (b) 180 (c) 220 (d) 2%
4. (a) Lower limit = 6 days, Upper limit = 10 days (b) 6 days (c) 8 days
(d) 98% (e) 84%

8. (a)

In the Sample	
Possible Number of Accounts	*Possible Proportion of Accounts*
0	0.00
1	0.25
2	0.50
3	0.75
4	1.00

(b)

In the Sample	
Possible Number of Accounts	*Possible Proportion of Accounts*
0	0.0
1	0.1
2	0.2
3	0.3
4	0.4
5	0.5
6	0.6
7	0.7
8	0.8
9	0.9
10	1.0

14. (a) 0.68 (b) 0.02 (c) 0.84 (d) 0.16 (e) 0.96
16. (a) 0.04 (b) 0.68 (c) 0.96
18. (a) 0.25 or less, 0.30, 0.35, 0.40, 0.45, 0.50, 0.55, 0.60, 0.65, 0.70, 0.75 or more (b) 0.50 (c) 0.74 (d) 0.26
20. 0.68, because the sampling distribution is normal and so the distribution in Figure 13.1 is applicable

Chapter 14

6. 0.05
8. (a) 0.98 (b) 7–13 days (c) 6 days (d) 10 days (e) 3 days (f) 0.98, 0.02
10. (a) 8–12 traffic accidents (b) 4 traffic accidents (c) 0.95 (d) 10 traffic accidents
12. (a) 25–35 mortgage loans written (b) 24–36 mortgage loans written (c) It would be wider because the maximum expected error would be larger.
14. (a) 0.98 (b) 0.528–0.752 (c) 0.112 (d) 0.224 (e) Confidence interval narrower, maximum expected error smaller
16. (a) 6,400 (b) 1,120 (c) 8,880–11,120
18. (a) 900 plants (b) 189 plants (c) 711–1089 plants
20. (a) 1250 peaches (b) 600–1900 peaches (c) 700–1800 peaches
22. (a) No, it does not satisfy the requirements. The maximum expected error is 2 instead of 5 and the confidence coefficient is 0.98 instead of 0.95. These differences indicate that the sample used was too large and so the survey cost was higher than it should have been.

(b) No. The maximum expected error is too large (10 instead of 5). This estimate is not useful to the health-study team.
(c) Yes. It is not to be expected that the maximum expected error will be exactly as specified, but it should be close (and 4 is close to 5). The confidence coefficient is 0.95 as required (this should always be true).

Chapter 15

2. (a) Parameter: population mean number of hours of light bulb life for company A
Statistic: mean life for a sample of bulbs
(b) Parameter: proportion of calls over a long period of time (the population of calls) which result in an order
Statistic: proportion in a sample of calls
(c) Parameter: mean weight gain per pig for a population of pigs
Statistic: mean weight gain for a sample of pigs
(d) Parameter: proportion of the population of students who enroll for this course over a period of time who are females
Statistic: proportion of a sample of students who enroll for this course who are females
(e) Parameter: difference between the proportion of college freshmen who drop out and the proportion of college sophomores who drop out in the populations of college freshmen and sophomores
Statistic: difference between the proportions for a sample of freshmen and a sample of sophomores
(f) Parameter: mean height of all plants of the given variety grown in the southern part of the state
Statistic: mean height of a sample of these plants.
(g) Parameter: Proportion of cases handled in the last two years which were handled out of court
Statistic: Proportion of these cases handled out of court based on a sample of cases.

8. (a) *Hypothesis to be tested.* Cure rate = 0.70
Alternative hypothesis. Cure rate is less than 0.70
(b) *Hypothesis to be tested.* Proportion out-of-shape = 0.20
Alternative hypothesis. Proportion is more than 0.20
(c) *Hypothesis to be tested.* Mean diameter = 2 inches
Alternative hypothesis. Mean diameter is more than 2 inches

10. Type I error: If hypothesis being tested is true, it is rejected, and alternative hypothesis is accepted
Type II error: If hypothesis being tested is false and it is accepted

16. (a) *Hypothesis to be tested.* Proportion of infected peaches in the carload lot = 0.05
Alternative hypothesis. Proportion is greater than 0.05
(b) Type I
(c) 0.01 is more suitable because it represents a small probability that the error will be made
(d) Type II
(e) 0.05 is more suitable because it represents a high probability of rejecting the claim if it is false

18. (a) 0.005 is more suitable because it would make the probability of accepting the claim if it is false very small
(b) 0.01 is more suitable because the probability of rejecting the claim if true would be smaller

20. (a) \$500 (b) Sample means of \$500 and above (c) Sample means below \$500
(d) 0.02 (e) 0.05

Chapter 16

2. *Hypothesis to be tested.* Mean = 120
Alternative hypothesis. Mean is less than 120

4. Region of acceptance: \$175.25–184.75
Region of rejection: less than \$175.25 and more than \$184.75

6. (a) *Hypothesis to be tested.* Mean = 20 days
Alternative hypothesis. Mean is less than 20 days
(b) One, because alternative hypothesis is one-sided

8. (a) *Hypothesis to be tested.* Mean = 20 days
Alternative hypothesis. Mean is not equal to 20 days
(b) Two, because the alternative hypothesis is two-sided
(c) Two, a lower-tail critical region made up of sample means below the lower-tail critical value and an upper-tail critical region made up of sample means higher than the upper-tail critical value
(d) Sample means equal to the two critical values and sample means between them
(e) 0.01, because this is the level of significance of the test

10. (a) *Hypothesis to be tested.* Proportion = 0.25
Alternative hypothesis. Proportion is less than 0.25
(b) 0.05, probability of rejecting the claim if it is true
(c) One, because the alternative hypothesis is one-sided

12. (a) *Hypothesis to be tested.* Proportion = 0.75
Alternative hypothesis. Proportion is not 0.75
(b) Two, alternative hypothesis is two-sided

14. (a) *Hypothesis to be tested.* Proportion = 0.40
Alternative hypothesis. Proportion is less than 0.40
(b) One-sided alternative hypothesis, lower-tail test

16. (a) 0.54
(b) Sampling proportions below 0.54
(c) Sampling distribution of a proportion
(d) 0.05
(e) Sample proportions equal to 0.54 or more
(f) Sample proportion = 250/600 = 0.417 which is below the critical value and so lies in the region of rejection. Therefore, reject the claim and conclude that the population proportion is less than 0.80.

18. (a) *Hypothesis to be tested.* Standard deviation = 20 books
Alternative hypothesis. Standard deviation is greater than 20 books
(b) One, because the alternative hypothesis is one-sided
(c) 0.05

20. (a) Upper-tail (b) Lower tail (c) Both tails

22. (a) Larger for 0.02, because the critical region includes 2% of possible sample proportions compared with 1% for the 0.01 level
(b) Larger for 0.05, because 5% of possible sample means are included compared with 0.5% for the 0.005 level

24. (a) This test has an upper-tail rejection region. If the significance level were reduced to 0.01, this region would get smaller and the critical value would have to increase to cut off a smaller tail area.

(b) Rejection region would get larger, so the critical value would have to decrease to cut off a larger tail area

26. (a) Lower-tail critical region would get smaller and so the critical value would become smaller

(b) Upper-tail region would get smaller and so the critical value would become larger

Chapter 17

2. Time reactions after drug administration, because there is no population of time reactions for subjects after drug administration

4. (a) The mean number of days is the same for the two periods.

(b) The mean for the last two years is larger than for the previous two years.

(c) The means are not the same.

6. *Hypothesis to be tested.* Proportions are the same
Alternative hypothesis. Proportions are not the same

8. Difference between the mean prices (2¢) falls in the region of acceptance. Conclude that the mean prices are the same.

Chapter 18

2. (a) 11.345 (b) Conclude that the proportions are not all equal.

6. Conclude that the characteristics are independent.

18. *Hypothesis to be tested.* The sales forces in the three branches are equally effective
Alternative hypothesis. They are not equally effective

Chapter 19

2. (a) Independent: annual gross sales
Dependent: average weekly wages

(b) Both

(c) Correlation analysis, the values of both variables are random

4. Correlation model, the values of both variables are random

6. Regression model, only the variable "number of days in court" has random values

10. Correlation model. You cannot evaluate the goodness of the correlation coefficient if the regression model is used.

Chapter 20

6. Slope and height of the regression line.

10. (a) -0.20, indicates that cost goes down 20¢ for each additional square foot added to living space

(b) Negative, lower number of square feet is associated with higher cost

(c) $200

(d) Predicted cost = 200 − 0.20 × square feet

(e) $140, $120

12. (a) Negative, lower family income is associated with greater number of visits to public health clinic
(b) -0.10, indicates that number of weekly visits decreases 0.10 for each dollar increase in family income
(c) 23 visits
(d) Predicted number of family visits $= 23 - 0.10 \times$ family weekly income
(e) 14, 3

14. (a) 0.20, indicates that the percentage of females in the labor force declines by 0.20 (percentage point) for each percentage point increase in the population living on farms
(b) 40.5% (c) 34.5%, 26.5%

Chapter 21

2. (a) 47 bushels per acre
(b)

	Bushels per Acre
Total deviation	20
Deviation due to regression	2
Unexplained deviation	18

4. (a) $1500
(b) Total deviation $= -\$2000$, deviation due to regression $= -\$2500$, unexplained deviation $= \$500$

6. Coefficient of determination $(0.86)^2 = 0.74$ indicates that 74% of the variation in one variable (say, crime rate) is associated with the variation in the other variable (based on the regression equation).

8. (a) 1500 (b) 0.75

10. Coefficient of determination $(-0.95)^2 = 0.90$ indicates that 90% of the variation in one variable is associated with the variation in the other variable (based on the regression equation). The negative sign means that an increase in one variable (say, price) is associated with a decrease in the other variable.

Chapter 22

10. 31.3 new loans. Prediction is very risky because 90 is quite a large extrapolation.

Appendix A

2. 2.44–3.06 inches. The fpc should be ignored because the sample size is a negligible proportion of the population size, so the fpc is practically equal to unity.

4. *Hypothesis to be tested.* Mean height = 8 centimeters
Alternative hypothesis. Mean height is not equal to 8 centimeters
Decision rule: If the sample mean height is either less than 7.233 centimeters or more than 8.767 centimeters, reject the hypothesis tested, accept the alternative hypothesis, and conclude that the population mean height is not 8 centimeters. Otherwise, conclude that the population mean height = 8 centimeters. The sample mean height (9.2 centimeters) falls in the critical region. Do not accept the claim.

6. *Hypothesis to be tested.* Mean running time for the population of high-school boys is the same for dry days and rainy days
Alternative hypothesis. Mean running time is not the same
Decision rule: If the mean of the (rainy day time-dry day time) differences for the sample of boys is either less than -3.50 minutes or more than 3.50 minutes, reject the null hypothesis, accept the alternative hypothesis, and conclude that mean running time is not the same for dry days and rainy days. Otherwise, conclude that the mean running times are the same.
The mean of sample differences (12 minutes) falls in the critical region. Conclude that mean running time is not the same for rainy days and dry days for the population of boys.

8. *Hypothesis to be tested.* The proportion of $+$(plus) differences (sample A $-$ sample B) is 0.5
Alternative hypothesis. The proportion of $+$(plus) differences is greater than 0.5
Decision rule: If the number of $+$(plus) differences for the sample of pairs is greater than 16.02 reject the null hypothesis, accept the alternative hypothesis, and conclude that time required for the drug to take effect is more for type X5 (used for sample A). Otherwise, conclude that time required is the same for both types.
The number of $+$(plus) differences in the sample $= 15$ and the lower limit of the 14.5–15.5 interval falls in the region of acceptance. Conclude that time required is the same for types X5 and X8 of the drug.

10. The same result is obtained. $U = 92$ falls in the region of acceptance. (In Problem 9, $U = 74$ also fell in the region of acceptance.) It does not matter which sample is called n_1.

12. 0.74

Index